The PLANET of the THINKING ANIMAL

SURVIVING THE 21ST CENTURY

*For Donald George Young and in memory of
Eileen Joyce Young*

The PLANET of the THINKING ANIMAL

SURVIVING THE 21ST CENTURY

TOR HUNDLOE

The Planet of the Thinking Animal: Surviving the 21st century

By Tor Hundloe

Published by JoJo Publishing

'Yarra's Edge'
2203/80 Lorimer Street
Docklands VIC 3008
Australia

Email: jo-media@bigpond.net.au or visit www.jojopublishing.com

National Library of Australia
Cataloguing-in-Publication data
                Hundloe, T. J. (Torstein John)
                The Planet of the Thinking Animal: Surviving the 21st century

                ISBN 9780980619300 (pbk.)

Edition: 1st ed.

                Sustainable development.
                Sustainable development--Scandinavia--Case studies.
                Sustainable living.
                Natural resources--Management.
                Energy industries--Management.
                Global warming.
                Environmentalism.

                338.927

Editor: Gill Smith
Designer / typesetter: Rob Ryan @ Z Design Media
Printed in China by Everbest Printing

CONTENTS

THE AUTHOR

Tor Hundloe is a pioneer of the global environment movement. His profound interest in nature commenced in his formative years living on a small farm in southern Queensland, in the Gold Coast hinterland. The farm was surrounded by rainforests. Today these forests are listed as the Gondwana World Heritage Area.

Tor had not commenced school before the Hundloe family headed for Norway and farm life in the 'Land of the Midnight Sun'. His grandparents' home was only a few hundred kilometres south of the Arctic Circle.

The contrast between Australia and Norway has stayed with Tor all his life. It is not only the difference in natural landscapes (rainforests and deserts in Australia, fjords and snowfields in Norway), but the different social and political cultures that have driven Tor's passion for sustainability science. The fact that people and places have common needs if they are to be healthy, yet there are vast differences around the world, has led Tor to explore as much of the world as possible in search of solutions for the long-term good of people and the other animals with which they share one small planet.

In 1974, Tor was the first Australian to seek election as a proto-Green in the national parliament. In 2003, he was made a Member of the Order of Australia for his contribution to environmental economics and related matters. In the same year he was awarded a Centenary Medal for education.

Today Tor Hundloe is a Professor at two universities. At Griffith University in Brisbane he leads a small team of researchers investigating the use of farm wastes to make bio-fuels (ethanol and bio-diesel). At Bond University on the Gold Coast he is working with another small team to develop one of the world's first undergraduate programs in new Sustainability Science, with the first students in this degree to be enrolled in 2010. Tor is also Emeritus Professor in Environmental Management at the University of Queensland.

The Planet of the Thinking Animal is Tor's third book. His second, *From Buddha to Bono: Seeking Sustainability*, was published in 2007 to critical acclaim, and is the recommended prequel to *The Planet of the Thinking Animal*.

ACKNOWLEDGMENTS

I commenced writing this book while I was employed as Professor of Environmental Management at The University of Queensland. I thank the university for its support during this period. I left The University of Queensland in 2006, after ten years, and took up the position of Professor, Environment School, Griffith University, where I completed the research and writing tasks.

As I write anything of substantial length by hand, I have engaged various people to type the manuscript. I thank Tara Cully, Michelle Wenner and Judy White for a great job in deciphering my scrawl. It tends to become particularly difficult to read when I write while eating a curry in my favourite Indian eatery, the Balti Café on Lygon Street, Carlton, Melbourne.

I must thank Charlotte Strong of JoJo Publishing for taking this book under her wing. Her eagerness to see the project through to its conclusion inspired me to keep my head down when other matters (both work and pleasure) could have intruded. Charlotte arranged for Gill Smith to be my editor. I appreciate the considerable work she put in to improve the book.

Finally, I was pleased when JoJo Publishing took me on as an author for the second time in a short period. Nothing surpasses the personal style of this small publishing firm.

INTRODUCTION

This book is set in numerous cultures in various countries, but has a common theme – how to survive the twenty-first century. Resilience. Sustainability. These two words catch the spirit of the early twenty-first century. After a century of environmental neglect – in fact, denial by some of the worst polluters in the world – we have awoken. Global warming is reality. It is no longer a theory. Practical people faced with the job of countering the threats of potentially dramatic climatic surprises, melting icecaps and sea-level rises are seeking solutions to lower and hold constant the emission of greenhouse gases. We will need to reduce them by 60 to 80 per cent in a short time – by the year 2050. Practical people will also have to deal with – adapt to, if you like – the adverse impacts that cannot be halted, whatever actions we take today. Sea-level rises resulting in flooding in low-lying countries are the most obvious impacts we will have to live with until we turn the greenhouse ship around. It is as cumbersome as a giant tanker, and as dangerous as one loaded with crude oil.

It has also dawned on us that, notwithstanding stable or even declining populations in the rich world, a demographic growth projectory means we must expect a 50 per cent increase in the world's population before it stabilises at nine billion plus within the next 30 to 40 years. Virtually all the population increase is going to occur in the poor countries, some of which are desperately poor. Three results are certain: greater poverty, increased degradation of local environments as the hungry search for their next meal, and even more carbon dioxide released into the global atmosphere.

There is a fourth fact to understand. The metaphor of China as an awakening sleeping giant is apt. We rejoice with those Chinese who have become middle class and can start to enjoy the goods and services that the world's middle class have taken for granted since the consumer revolution took off in the 1960s. In concert with the pleasure we feel for these hard-working and fortunate Chinese, we worry a lot about the smokestack pollution that goes hand in hand with their economic achievements. They and the even poorer Indians, Indonesians and the desperately poor Bangladeshis and Pakistanis – not to mention many sub-Saharan Africans – need and deserve economic growth. Don't let any green advocate deny this.

Are the world's ecosystems resilient enough to absorb the shocks we are unleashing upon them? We are not sure, but we should be very worried. Ecological footprinting – calculating the amount of land and resources required per person – suggests that a world population of people with middle-class lifestyles would require us to discover and then settle another two planet earths!

There is, therefore, an urgent need to understand in considerable detail what we are doing to the globe's ecosystems. We need timelines, loud warning signals well before we overshoot thresholds and, of course, solutions. To make progress we must seriously ramp up ecological science. We need mega-scientific projects led by outstanding individuals. At present, far too much scientific effort in ecology goes into providing (too often ambiguous, but nevertheless strongly argued) options on local land-use disputes or media-sexy projects. Iconic fauna, coral reefs and rainforests trump the lowly earthworm, agricultural soils and grasslands. We need ever more precise and rigorous studies of the Limits to Growth kind. The first 'Limits' study was published in 1972, the second in 1992 and the third in 2004. The 'Limits' study was a major computer-modelling exercise, tracing through to their extremely unpleasant conclusions the effects of ever-increasing human population, unlimited economic growth, the depletion of finite resources and much more. We know much more now and the urgency exists for frequent studies of this kind. We must make best use of scientific breakthroughs as they occur.

STRONG SOCIAL CAPITAL

To fully understand the problem, we must look further than the science to the important role played by strong social capital in forming a desirable sustainable society. A free press helps greatly; so do good laws that are enforced. The most important prerequisite is genuine democracy. The longest lasting democratic institution in the world is Iceland's Althing.[1] This legislative assembly was established in AD 930, not long after the country was settled. For a brief moment consider this tiny country (both in size and population) which, in 2007, ranked at the top of the United Nation's human development index (HDI). This was its first time on the top of the world's list of overall best performers. Before Iceland's ascendancy, Norway was in top position from 2001 to 2006. Only one other country,

Canada, has been on the top for more times (ten) than Norway. Maybe the case of Iceland, and, of course, those of Norway and Canada, offer solutions – or at least answers – to some of our questions.

From an environmental perspective, Iceland performs well above its weight. Consider some pertinent facts. Notwithstanding Iceland's lack of arable land (only a tiny 0.07 per cent of total land), the country's advantage is its enormous supply of geothermal energy and hydropower. Combined, these two renewable sources of energy provide 99 per cent of the country's electricity. As a consequence, the country utilises a relatively insignificant amount of fossil fuels for transport purposes. And the end for fossil fuels is in sight in Iceland. Hydrogen-powered buses ply the streets of Reykjavik, its capital. Iceland aims to be the first country in the world to walk away from the oil age.

If a tiny frozen country like Iceland can be a leading force for sustainability, the rest of us (the larger resource-rich countries) must be able to do much, much more. But we must be clear about what our goal is.

OUR GOAL

We are interested in ecosystem resilience because we wish to achieve, and then sustain, a good life for humans, and for the other animals with which we share the planet. We desire to sustain good things. We are not engaged in a simple Darwinian struggle for survival. We want more than simple existence, and have proven that we can achieve more when we apply science to human needs. We have virtually doubled the lifespan of middle-class people in a mere 100 years. We have tools and toys of great variety to assist us in our daily work and recreation.

Our understanding of what we want and how to obtain our goals is poor. This is a result, by and large, of our economic, sociological and political science endeavours being left to languish in the dingy corridors of poorly funded and poorly staffed university departments for the best part of a 100 years. Our social science has withered on the vine. We made excellent progress in the Enlightenment, and only recently a resurgence (a new Enlightenment) has occurred.

WE LACK UNDERSTANDING OF HUMAN BEHAVIOUR

As the social science disciplines have fragmented into ever-narrower sub-disciplines – and adopted without question the most basic behaviourist theories of human nature – we have lost the inquisitiveness that drove the Enlightenment thinkers. Re-read Adam Smith, David Ricardo, Thomas Malthus, Karl Marx and John Stuart Mill to discover what social science can be about.

Since we have given the ecologists a mammoth task in determining the limits of the globe's carrying capacity, let us give an equally large one to the economists and social scientists. Part of their work will need to be done in concert with the ecologists, something that will become obvious as we unravel our challenges. The most pressing task the ecologists have is to ascertain how many humans can inhabit the globe (we are not going to find another two) on a sustainable basis, with lifestyles that meet our desire to be happy. How do we get our fellow humans to work together to limit our numbers, to limit our material demands and recognise that we can be fulfilled in a world which is not fashioned to the endless pursuit of out-consuming 'the Joneses'?

What little we know

Let us consider some things we already know. We already know that the further down the food chain we eat, the more of us can be fed. Likewise, we already know that the more we reuse and/or recycle materials, particularly those things we presently call wastes, the more of us the globe can sustain. Furthermore, we know that fossil fuel limitations – as well as the finite supply of other non-renewables – will bring our civilization to a crashing halt if we don't quickly adopt the technologies that utilise renewable energy sources.

We know much about physical limits and we accept the findings of biologists and botanists, and yet when it comes to changing behaviour so that we deliver on the common human desire for the good life, we know very little. We struggle to understand how to do it. The evidence is stark. The twentieth century was truly horrible in the waste of human life in world wars, civil wars and genocide. We were – and are – very cavalier when it comes to human life. The carnage on the world's roads would be unthinkable in a world where human health and happiness mattered. The drive to extend our personalities via shiny chrome, artificial leather upholstering and instant music is extremely difficult to explain. Our inability to make the motor car a source of good (which clearly it could be) illustrates our very poor understanding of human behaviour and how to change it. Each and everyone of us who drives a car believes that we gain more benefits than costs from this action – although we worry about our vehicle being stolen, damaged in a shopping-centre car park, and breaking down, and (if we are at all rational) we worry that every other driver on the road is a potential idiot. That only a few are idiots is enough to kill and maim more people than wars do. We abhor war; we turn a blind eye to the carnage on the roads.

How little we care

Take another dysfunctional aspect of human life. We know how to grow enough food to feed the world's entire population, yet we don't know how to distribute it so that no-one starves. Near to one billion people do starve.

We know how to mine the world's mineral wealth (oil, coal, iron, bauxite), but not how to distribute the profits fairly and how to reinvest the profits to achieve a truly sustainable income. These are matters few comprehend or, if they do, care about. How many desperately poor countries have produced enormous wealth for a corrupt dictatorial leader to carry it off (as they always do) never to be seen again or simply just squandered wealth as Nauru did with its phosphorus riches? It is not only the Third World countries that don't get these policies right. There are rich countries squandering their wealth. Any country that depletes its non-renewable resources without making compensatory investments is guilty of squander. We know the answers to some of these economic and social problems – ones that are denying us a good, sustainable life – but a number of important problems remain a mystery.

This book takes as its starting point the reality that individuals, and collections of individuals in nation states, are hesitant to admit that they do not have the answers. Let us admit ignorance while accelerating our search for answers. We can learn from the mistakes of others and, more importantly, we can learn from the success of others. We can learn from the great variety of experiments that are underway across the world. In the following chapters we investigate failures and successes in the search for a better, more environmentally sustainable world. Some countries seem to be better than others in searching for a sustainable future. There are many stories to be told and are worthwhile following. Let us start the investigation: a case-by-case study around the globe. Each of our case studies will help us to move forward.

SUSTAINABLE USE OF WATER

We commence in Chapter 1 with water to illustrate the challenges of meeting the key principles of sustainable development. Humans are mainly water. We die within days without it. Due to trans-boundary flows, we fight over it. The use and overuse of water epitomises one of the fundamental problems of working towards sustainable development. How do we ensure people get a 'fair share' of clean water, and how do we ensure enough clean water is in the right places at the right times for future generations? The problem is not that there isn't enough to go around. It is purely a matter of allocation of clean water in space and time, given the number and distribution of people and other animals on the earth. Fewer of us means less water required at any point in time.

As we do with food, we can use the price mechanism to allocate water, but would this approach deliver the outcomes we desire if our aim is to ensure a fair share of water for all? The price mechanism does not result in food being provided on a needs basis. If it did, we would never again see a malnourished, potbellied, wide-eyed African child clinging to life until the next aid package arrives – the package that comes too late for too many.

Why should we expect the price mechanism to do better with water? Because fresh water is so critical to life, should it not be a resource owned and allocated by the community working to a 'golden rule' principle – do unto others as you would want them to do unto you? That is, ensure your fellow humans get enough water to maintain a healthy life. We must face the matter of inequities involved in using the market – as we know the market at present – to provide water and food for all. We will need to consider how to modify the market's operation so as to meet human aspirations. We either do this or install a dictatorial system by which to allocate water, food and much more!

Who owns water?

Back to water. Today we claim the water that falls on 'our country' (read, our piece of land) and, hence, believe we have the right to do as we like with it. If we are upstream at the top of the catchment, we have the potential to have a major impact on the quantity and quality of water downstream. First in, best dressed!

This raises a very basic question. Is climatic good fortune fair? Some of us are born in wet tropical rainforest areas, others in mild Mediterranean climates, others in windswept deserts and yet others in frozen Arctic icefields. None of us asked to be born where we are. For the majority of humankind, where they are born determines their future – only the affluent middle class is free of geographical, economic and cultural determinism.

Water belongs to all – or to no-one

The fact that some countries are nearly 90 per cent desert is not fair. (If the same countries are flush with oil, this also is not fair – except that it compensates for the lack of arable land and fresh water.) Much that is determined by nature (or the gods) is not fair in the world. Water is too critical for life to be denied to any one individual. We need to determine how to override the initial (natural) allocation of water so to make it fair, or at least fairer.

Given that the water drunk by a Chinese farmer 2000 years ago could be what your cup of coffee contains, water allocation between countries is going to require global oversight. The point is that the same water that belonged to a Chinese farmer 2000 years now belongs to you – momentarily. Water belongs to all or to no-one! Just as we have come to recognise the atmosphere as a commons – no-one owns the air we breathe – and that the world's climate is a global commons – hence the Kyoto Protocol – so we are going to have to change our mindset about water.

In the 1990s, with the fashion of infrastructure privatisation in much of the Western world, community after community lost control of their water. Without asking permission, governments sold the people's water to private corporations. Was this a good or bad thing? Water is a common-property resource. Yet has not history shown us the advantages of private property? It would seem that it is not neo-conservative propaganda to claim that privatised farmers' plots in China have been much more successful in providing food to the masses than were the communes. In the same vein there are virtually unlimited examples of the advantages of providing people with an economic incentive to use something – anything – wisely. Private ownership is, so it seems, part of if not all the incentive for better decisions on water use. But before we sign off – as our politicians did in the 1990s – on selling off our water and whatever else caught the eye of the global entrepreneurs, let us consider the reverse side of the coin.

Privatise water?

We have found throughout history that there are circumstances in which cooperative behaviour in utilising a resource is the best economic solution. Today the benefits of cooperative behaviour underpin the United Nations Law of the Sea and the numerous successful policies to control trans-boundary pollution in Europe and North America. Let us attempt to make sense of water allocations on the principles of cooperation.

The following thought process helps us. First, water is not a farmer's block of land. It is not stationary, to be appropriated by whoever holds it. Water is in a continual cycle, while land is immobile. The person occupying land is the beneficiary of the land. Water is never in one place long enough to be owned. You cannot stand in the same stream twice, as has been famously stated. If you think otherwise, consider where that glass of water you drank yesterday – when for a few seconds it was yours – is now. To attempt to apply the principles of land ownership to water ownership is nonsensical, not necessarily for economic reasons, which can, surprisingly, neglect real-world physics, chemistry and biology, but because of the laws of nature. A global water protocol recognising the unique characteristic of water is urgently needed. Sorry to say, this seems to be beyond human competency given our level of intellectual development. We cannot remain at this level of ignorance.

FROM WATER TO FISH

In Chapter 2 we progress from water to what lives in water. The fishing industry, or the seafood industry to be more inclusive, has been the mainstay of humans from our earliest days. Human settlements developed along rivers and on the coast. These environments were – and are – some of the easiest to be exploited by hunter-gatherers. Of course we did not think of fishing as an industry when we were hunters and gatherers.

Today much of the world relies on seafood for protein. It is harvested, by and large, by profit-seeking fishers and fish farmers. Yet, in the poor countries, people still gain their seafood as hunters and gatherers. In the rich countries people called recreational fishers or anglers supplement their diet by their fish catches.

The world's oceans are fully exploited – some are over-exploited – because of what scientists call the 'tragedy of the commons'. The tragedy of the commons is one of the most fundamental economic and environmental problems we have faced since human populations exploded following the success of the Industrial Revolution, which began in the United Kingdom in the late eighteenth century. A commons was an area of land or water to which no-one could lay claim; as a consequence much over-fishing occurred in the oceans.

First in, best dressed

In the simplest terms, the rule was 'first in, best dressed' – catch the fish before someone else does – a race that would see massive waste of fuel, material and labour in a no-win economic pursuit as fewer and fewer fish were caught, until virtually none can be caught. Many a fisherman became bankrupt in this perfectly 'rational' approach – rational if you overlook the economic consequences of exploiting a common resource to which people have free access.

Without a law to deal with sharing common property, it is completely rational to take as much as you can before someone else does. As recently as the mid 1970s, nations went to war over access to fishing grounds because there were no universally recognised rules about sharing the commons. Recall the 'cod wars' between tiny Iceland and the UK – a David and Goliath battle with the proverbial happy ending. It is a rather esoteric point to make but it is important enough to be made – what is irrational from an economic perspective does not necessarily have to be from a private commercial perspective. In other words, the individual fisher can act rationally by setting out to maximise his returns, yet in doing so work against his goal and that of all other fishers.

The cod wars

Natural resources are usually very valuable (in some cases necessary for life) and it only takes a media campaign based on 'national rights' (or worse still, religion) to stir people to kill for food, water or oil. The calm, cool, rational Icelandic people were not about to become barbarians when faced with a foreign threat to their major industry. During the cod wars, the Icelanders deliberately aimed their patrol-boat guns anywhere except at humans (that is, at Englishmen). By design, no-one got killed. Here was a 'civil' war – in an unusual use of the term. As a consequence of Icelandic perseverance and resistance more than 30 years ago there are now exclusive economic zones around maritime countries extending out 200 miles (320 kilometres). However, in some parts of the globe these fishing zones came too late, as the oceans were already fully fished or over-fished.

How do we feed the masses?

Where to next to feed the extra three-plus billion people who are joining us as the population increases in the next 40 years? Unless we change our diets, seafood will be in very strong demand, prices for seafood will increase and much pressure will be put on our oceans as well as on fish farms and aquaculture. The details of this story are told in a later chapter.

KEEPING WARM OR COOL

We drink, we eat and we need to cook our food, and we need to keep warm and cool. This requires more than the energy our bodies generate. We require an outside source of energy. This has been the case throughout recorded history for the tribes of the Arctic: animal oil helped to keep them warm and provided light throughout the sunless months for half of the year.

As we explore in Chapter 3, our present-day technologies – based on mechanised transport and electricity generation – can only continue indefinitely if we utilise renewable fuels. This is a rude shock. From the time that the British navy, in the early twentieth century, made the decision to fuel its fighting vessels with oil, the future was determined – until now. Crude oil is close to its peak production and we will soon be using more than 4.5 litres of oil to recover 4.5 litres from nature. Coal is still abundant but demand is rapidly increasing and climate change will eventually halt its use well before it runs out. Stones remained after the end of the Stone Age – so will it be for the oil age and then the coal age.

The search for new fuels

The search for alternative energy sources is not new. We only think it is because, for the first time in history, our lifestyles are threatened by the prospect of the end of the use of oil and serious climate change. Hardwired human inventiveness leading to technological breakthroughs has been the basic driver in the search for new fuels. When humans did all physical work by muscle power, we hardly needed economic incentives to search for fuels – our muscles merely demanded relief. To search for and invest in something new is part of the human disposition. Notwithstanding our hardwired attributes, economic imperatives can accelerate (or retard in some cases) the process of innovation. A successful breakthrough in finding new energy sources will have dramatic consequences on the nature of an economy and society.

When the steam engine was invented, the demand for coal shot up dramatically. When the internal combustion engine was invented, oil demand escalated and continued to increase as more and more cars were produced and bought. Ironically, the fuel source for our cars could have been peanuts, as this was the fuel of choice for the original diesel engine. Imagine the world today if peanuts fuelled our cars, trains and planes, drove our electricity generators and heated (or cooled) our homes. This is a wonderful topic for a 'what if' book.

We must move on, recognising that the discovery of electricity and its utilisation in household and industrial applications has changed the world dramatically. Coal, oil, natural gas, uranium and, in a few countries, falling water have become the favoured sources of energy. Peanuts remain for eating – and even this is a problem for some who have discovered that they have a potentially fatal allergy to this nut.

Pollution drives our enthusiasm

Pollution has come to play a catalytic role in the search for new energy sources. The London smog of 1952, in which about 5000 people died, was a wake-up call for those who didn't understand what economists call 'externalities' (the cost of pollution which only the victim pays). As a consequence of the 1952 catastrophe, the wealthy countries began to regulate emissions from household chimneys, from industrial furnaces and smelters, and from motor vehicles. More than 50 years later, poor countries believe (wrongly) that they cannot afford to control pollution – they are afraid that their economies will suffer and their people starve. Selling cheap T-shirts to the West is the only way forward – so they believe.

The different worlds

The situation is different in the rich countries. It is inevitable that air quality will continue to improve in the Organisation for Economic Co-operation and Development (OECD) countries as ever more stringent laws are put in place. The concept of 'zero emission' vehicles has already led to increased sales of hybrid electric vehicles. The wheel has turned – but maybe not in the USA as gas (petrol) remains inexpensive and the culture favours the ownership of big, fuel-inefficient vehicles.

Iceland aims to be the first country in the world to have a hydrogen-fuelled vehicle fleet. When this is achieved no fossil fuels will be used in Iceland. Sweden has set a target to be rid of fossil-fuel-powered vehicles in the next decade, and the possibility for a similar outcome exists in the South Pacific if Fijian sugar cane is converted to ethanol production. There is a range of possibilities for replacing fossil fuels in many countries if the political will exists and sound economic policies are allowed to determine outcomes.

Acid rain leads to Stockholm

What localised pollution did for domestic environmental laws in any number of countries, trans-boundary pollution did for global environmentalism. More than 40 years ago, it was discovered that acid rain from coal-fired power stations was causing severe damage to forests and lakes in Scandinavia and parts of Canada. The power stations were often situated in countries other than the affected ones, such as Germany and the USA. This was the background to the first UN conference on the environment, the Stockholm Conference in 1972, which was very much about the First World coming to recognise its environmental problems – as experienced by its citizens and its industries. The developing countries – and some not developing – were, by and large, neglected at the conference. The time was not right for a genuine global perspective.

Positive responses

Today, 30-plus years later, there are both domestic and international regulations governing the emissions of the oxides of sulfur and nitrogen (the cause of acid rain). As a consequence of these bans, there have been significant technological changes in coal-fired power stations, in particular de-sulfurisation at the flue. Recently, natural gas, with a lower carbon content than coal, has become the preferred fuel in countries attempting to reduce greenhouse gases.

It was not only air pollution that was targeted in 1972. The Stockholm Conference also led to serious attention being paid to water pollution, including trans-boundary water pollution in Europe where rivers travel through country after country then away to the sea.

The role of economics

Finally, we should note that economics has played, and plays, a critical role in the uptake and growth in demand for new fuels. The more people who are willing to purchase a new energy technology, the less its price, and as a result even more people will purchase the new technology. In other words, economies of scale are extremely important until they are eventually exhausted. Today, given the rapid technological advances we have made, the lack of economies of scale is the major constraint to the uptake of non-polluting fuels. It would take no more than the majority of consumers in one of the three large markets (the European Union, the USA or Japan) to adopt a new fuel for their motor vehicles for that fuel to become more than competitive with oil. It is as simple as that. There is no law of economics that favours fossil fuels.

The past constrains us

The difficulties we face today are that, as motor vehicle manufacturers, car owners, urban planners and freeway engineers, we made decisions in the past – when there were far fewer of us – that favoured fossil-fuel-powered transport and we are locked in to it. No-one was thinking about any fuel other than petrol or diesel in those days. It is going to be quite costly to build the infrastructure for distribution of new fuels. Imagine that petrol (gas) stations have been replaced by businesses that recharge car batteries or supply bio-fuels or fuel cells. That is what is demanded of us. Make the wrong decision this time and history will not treat us as favourably as it is treating our fossil-fuel forefathers. Our decision today has to be what type of energy economy we want in 2050 – the approximate date of the end of the fossil-fuel economy.

On the horizon

Hydrogen power is now on the horizon. It is the most abundant element in the universe. The hydrogen economy will mean the end of our concern with carbon-based greenhouse gases. The hydrogen fuel cell is an electromechanical device that combines hydrogen and oxygen to produce electricity and water. The devices are not new. They were first used in the US space program and more recently in submarines and army jeeps. Some good comes out of military expenditure!

At present, natural gas is the most common source of hydrogen. In the long run it will be renewable energy resources that will produce hydrogen at the least cost. There are various possibilities: electrolysis of water can convert solar and wind energy into hydrogen; biomass can be gasified to produce hydrogen. However, there is much research territory to cover before we arrive at the hydrogen economy and let us not jump into an utopian solution. In a later chapter we consider various alternatives.

PUT ASIDE OUR TROUBLES AND TRAVEL

Let us leave fuels to consider, in Chapter 4, one of the major uses of them: personal transport for holidays. For the world's ever-growing middle class, recreation and holidaying has come to mean travel. Travel – what we call 'tourism' – illustrates the challenges and opportunities involved in searching for sustainability. Travel is no longer the privilege of the affluent Westerner. It is no longer a positional good (something to do to show off one's wealth) but a middle-class 'right'. We expect to be able to travel to places far away, if we desire, for our annual holidays. Northern Europeans flock to Spain. If young, they are likely to end up in Australia as backpackers. The Japanese tour group can be found anywhere in the world – not that long ago only a few tried-and-true destinations counted. The middle class is rapidly expanding in the developing Asian countries: for example, Indonesia has a middle class the size of the Australian population of about 20 million and China's middle class is as large as the entire population of the USA and is growing. The economic success of the Asian middle class should not be overlooked. We should celebrate the fact that these people are no longer poor – everyone deserves a good, long and healthy life. Far too many people (hundreds of millions in countries we have just mentioned) have little prospect of that in the short term.

Tourism, as the world's largest employer of people, picks itself as having a significant role to play in furthering the income-earning aspirations of the poor. If poor nations can present their unique cultures and natural environments to paying visitors, why not? Through this industry sustainable incomes can be earned, rather than one-off payments from clear-felling forests or any number of environmentally destructive industries that the poor can be induced to engage in.

If only it was as simple as adopting tourism as the saviour of environments and the payer of bills. For all the good that results from tourism, examples from around the world show that the downside of tourism can potentially outweigh the positives. And there are numerous negatives to tourism. If it is not well managed, this industry can cause serious environmental degradation and resource depletion, and enable criminal activity and the loss of local culture and social capital. And we cannot overlook the fact that the climate change debate has finally caught up with international travel, most of which is undertaken by aircraft burning fossil fuels. While international travel is a minor contributor to greenhouse gases, its very existence demands attention.

Overall, if done properly, tourism is a means of good. The more tourists learn about nature, the more likely they are to become conservationists. The more people of different skin colours, creeds or cultures mix, the more likely they will think in terms of a 'common future' and think of themselves as citizens of the world rather than Americans, Chinese or Germans. Our challenge is to get people to experience nature and other cultures without causing environmental or social damage in the process.

After Chapter 4, the book changes focus from sustainability issues to certain countries or groups of countries and their particular sustainability problems and solutions. There are seven case studies, each with a sustainability problem (or problems) requiring solutions – sooner rather than later. We explore solutions to particular problems – it is not all negative. Each case study illustrates aspects of politics, economics or culture that either frustrate or facilitate progress towards sustainable development.

THE FUTURE OF FIJI IS TIED TO ITS PAST

The first case study, in Chapter 5, focuses on Fiji. This chapter is about the trials and tribulations that most small island nation states face, such as those in the South Pacific, which is spattered with small (some tiny) nation states and the Caribbean, as yet better known for the cricket performances by the West Indies team than its handling of environmental issues.

Much has been written about the unique biogeography of islands – think Galapagos and Charles Darwin's discoveries and where they led. Darwin proved that we can learn much as ecologists from the specific populations on an island. Yet the human dimension of island life and the specific common problems have not been as rigorously studied and documented. The history of an island's settlement, as much as its geographic isolation, provides grist for fascinating sustainability stories. Islands, simply due to their separation from the continents, tend to have a history of discovery and then waves of re-discovery. Continents are subject to a history of invasion – and often settlement – of outsiders. There is not a sense of discovery (except for the first invaders), rather one of incoming tides of other (continental) people. On the other hand, islands are too small and have too few people and too narrow an ethnic and cultural base to absorb with ease the next influx of foreigners. The 'invasion' of islands results, inevitably, in social conflicts, as the new arrivals compete with the earlier settlers for resources. We will come to understand this problem in our analysis of Fiji.

No two islands are alike

No two islands are alike. Some are absolutely fantastic transport hubs, as in the case of Singapore. Others are frozen outposts of Western Europe, such as Iceland, which is nevertheless the world's highest-ranking developed country on the HDI. Others have populations too small, resources too scarce and opportunities too limited to survive as nation states – take your pick of a number of tiny Pacific Ocean island mini-states.

Most islands are either scattered around the vast Pacific Ocean or concentrated in the Caribbean. By and large these islands are neither rich nor desperately poor but they do have serious cost imposts caused by distance from both import and export markets. The cost of transporting fuel to the islands is a prime example of a geographical disadvantage.

Then there are simple domestic matters that impact on island life. With limited land, where does the waste go as the populations grow and lifestyles come to replicate those of the wasteful West? What of overpopulation? Because they are developing countries, the islands have the developing countries' propensity to produce larger families than in the developed West. Children are future workers – the only factor of production the poor have. Yet opportunities for children to acquire a good education are limited. The unemployed youth of the Solomon Islands (to cite one example) become – ever too readily – the foot soldiers for the next uprising and coup.

Are there too many tourists to these alluring island holiday destinations? Does not the allure of an island holiday present an opportunity for governments and businesses to promote their island as an ecotourism destination? If so, at what cost? Maybe – possibly – the benefits outweigh the costs. These are the types of questions that face island states.

The Fijian example

There is a unique Fijian response to such questions. And yet there is more to the problems, and opportunities, facing this 'giant' of the South Pacific island nations. Fiji has a specific political problem: serious ethnic differences that affect its political life and economic future. The core problem is a dispute over resources, masked by ethnicity. That issues of ethnicity should arise in a country with Fiji's mixed cultural history illustrates just how insidious a blight racism can be on susceptible societies.

The intriguing fact is that millennia ago Fiji was the cultural, social and religious melting pot of the South Pacific, as successive waves of people moved from South-East Asia into the Pacific. Each new wave confronted an already settled society. Each wave were invaders and when they came to dominate the existing society, they were colonisers. That is forgotten today and their ancestors are happy to be simply Fijian, even though there are distinctive Melanesian and Polynesian people in Fiji. Both ethnic groups, although quite different in appearance, are today's Fijians. The more recent arrival around 100 years ago of other people (from India) is the root of the present ethnic problem.

Will ethnic differences fade away?

Must millennia pass before ethnic differences dissipate and eventually disappear? Are the last people to arrive – whether invaders, colonisers or simply immigrants – inevitably discriminated against? Do they come to dominate in due course? Is there a workable and preferable outcome based on the values of common humanity? Why shouldn't the various ethnic groups mix, intermarry and melt into a new group, a new society? These are the issues central to the future of Fiji – and other countries facing ethnic conflict.

Here we are dealing with the sustainability of a society. This is just as important as economic and ecological sustainability – without social stability, these will not get a look in. Time and time again in the near 40-year period of Fijian independence, no sooner has the country been on track to a sustainable future (based on tourism and manufacture of clothing and other products) than a coup based on an ethnic divide has occurred – the most recent in 2006 – and the Fijian society has gone backwards. There have been only losers!

OIL RICH, PEOPLE POOR

The next case study, in Chapter 6, deals with a completely different type of country, and one with vastly different problems. It is a small Arab country, rich in oil and forward-looking – in economic terms at least. It is the United Arab Emirates (UAE). The most frequently asked question in the UAE is what happens when the oil runs out? All countries dependent on exporting a non-renewable resource should ask this question. The UAE has a strategy for this inevitability.

The strategy is starkly evident in Dubai, a world-class shopping centre and the major trading port of the Middle East region – another Singapore if you like. For a sustainable future, those who control the oil wealth in any country need to be wisely investing a considerable amount of the profits. This they are doing in the UAE. As the country is not far from being an absolute monarchy, making wise (or unwise) decisions is easy.

The UAE is not all flash shopping centres and port facilities. It has its challenges: water is incredibly scarce. It is a desert country, where the sand storms, when they come, cover the sparkling cities of Dubai and Abu Dhabi like a funeral shawl. The air is too thick to breathe and only the escape to an air-conditioned shopping centre or hotel provides relief. Fortunately there are many of these, and none are overcrowded. When the dust storms come, only foreign construction workers remain out of doors, meeting the schedule to build the next iconic high rise.

Without air-conditioning the summer is unbearable in the UAE. The sea water is unswimmably hot, yet locals bathe in it, the women fully clothed in this still rather strict Muslim country. In some emirates – for example, Sharjah – alcohol is banned. Foreigners can only hope that their hotels have cooled the water in their swimming pools, but cooled swimming pools are hard to find. There is a technical solution to the lack of water: desalination is very affordable in the UAE, given its oil revenue.

What other country is able to – for the four quarters of an Australian Rules football game – become a de facto part of Australia? In February 2008, two famous Australian football teams played in the official pre-season competition in Dubai. One cannot imagine that happening in Ireland, a country with genuine claims to be the source of the Australian game.

A society of foreign workers

There is a reliance on foreign workers in the UAE. They to do most of the work from the basic manual construction jobs on the ever-expanding high rises to managing the five star hotels. Only one-fifth of the total population of the UAE are Emiratis; the remainder are 'guest workers' from South and South-East Asia, plus some from Egypt and a few Europeans. Can this arrangement persist indefinitely? The Emiratis don't want it to – or so they say. There are important psychological benefits for local people in not having to rely on outsiders. When foreigners are needed to undertake the more highly skilled tasks, the local people feel insecure and even like second-class citizens in their own country. Do these factors outweigh the benefits of an idle lifestyle? Only time will tell.

There are other small oil-rich countries in the Middle East and one in South-East Asia – Brunei – that share many of the same opportunities and problems with the UAE. Have they realistic post-oil strategies? What are their plans to replace guest workers with 'nationals', as they tend to call themselves? Will they, like the UAE, invest much of their oil profits in sustainable industries or waste it on consumption – and ultimately return to the poverty from which oil delivered them?

CRASHING THROUGH IN RUSSIA

In Chapter 7, the next case study tells the sorry story of Russia's attempted crash through to capitalism. With the very sudden demise of the Soviet Union in 1991, Russia and the newly independent Eastern European nations had a choice: move gradually to build democratic institutions fitting a social democratic (or, if preferred, democratic socialist) economy along the lines of the Scandinavian countries, or privatise holus bolus and wait for a new economy and new political situation to emerge. The latter route was taken and Russian society went backwards. Much of what was undone in that period will have to be fixed before there is real light at the end of the tunnel. A middle class is being established on the basis of the country's enormous oil wealth, but the proud Russians of the past are still few and far between.

Russia was not the only so-called communist nation to undergo rapid transition. All the European offshoots of the Soviet empire signed up to a new future based on capitalism. How have they faired? We are fortunate to have available considerable attitudinal data gathered from samples of citizens in these countries. The data highlights the success, or otherwise, of the opening up of these economies to the global economy. Overall, the picture is one of a successful political and economic evolution. The best performers in Eastern Europe have already joined the European Union; others are knocking on the door. However, as the chapter shows, some states – in particular, Russia – went backwards for a considerable period of time.

Boom town

In Asia, China is literally booming and Vietnam is steadfastly growing. The attitudinal data referred to above gives us an insight into one of the great experiments in human society: the remaking of China (and to a much lesser extent Vietnam) as capitalist societies – overseen by so-called communist parties. Such societal experiments in the history of human society are few and far between and we would be foolish not to make the most of any lessons.

It is not only the economic and political consequences of the transition that are interesting. Today we are concerned with much more. The old Soviet Union (as with communist countries generally) took virtually no interest in the environment. The lack of power of environmental lobby groups allowed governments do very much as they pleased. And they pleased to sanction short-term production of material goods at the long-term expense of nature and society: dump it – even nuclear waste – in the sea! Russia has not got the surplus funds it needs to clean up the environmental mistakes of the Soviet era (and ones it is presently creating due to its continuing laissez-faire attitude). But it is not only in Russia and the old Soviet Union that environmental vandalism is second nature.

BOOM OR DOOM

China, because of the overbearing power and control of its so-called communist government, still has to do much to clean its heavily polluted rivers and smog-filled skies. In Chapter 8, we review how little is being done at present as rapid capitalist development is resulting in enormous levels of pollution. Of course, China tried to convince visitors to the 2008 Olympic Games that it is a 'green' country. However, there is probably a generational gap (that is, 25 years) before its pollution prevention policies will be widespread and a benefit to Chinese society in general. I hope I am wrong on this count. China not only faces a massive pollution clean-up bill but also the just as urgent need to bring one-third of a billion desperately poor people into the new middle-class China.

The new middle class

Whether you live in New York or Auckland, if you have a role in promoting sustainable development – and we all do – you cannot but focus on China, then India. The future development of these two countries will drive the global search for sustainability. It is purely a matter of numbers. We are witnessing a middle-class revolution in these two countries – of enormous magnitude and at great speed. The Chinese and Indian middle classes want the middle-class lifestyles of the West, and we will deliver this to them by purchasing the goods that flood out of their factories into our shops and onto our bodies as cheap clothing, into our homes as cheap white goods, into our factories and farms as tools and machines, and into our offices as computers. It is a win for the new middle class in these countries (they get good wages); it is a win for Western consumers (we get inexpensive goods); it is a loss for the environment and a loss for future generations. Unless …

This makes it seem a simple one-way process. It is not. There is not one Asia. The rural poor in the two giants, China and India, outnumber the rest of the world's poor. There are hundreds of millions – in fact, well over a billion – of poor in these two countries. They await crumbs from their fellow citizens, whose children eat in McDonalds whenever they please. The great irony (the greatest in the present era) is that a China made on Marxist ideology is in line to be like another class-ridden USA. The Chinese rich will find it easier than the eighteenth- and nineteenth-century American robber barons, because in China the State is on their side. China was never communist and won't be in the future. We who thought they were communist were conned by their rhetoric and propaganda. A peasant society cannot leap over capitalism and become communist.

An ordinary life in Asia

Focus on the other nations of Asia, and one finds a mixed bag of development. Some countries are still very poor, while others aim to industrialise in the next 20 years. A very ordinary life in Asia is one of stoic acceptance of small changes, some positive, some negative. The stoicism of the under-classes is to be admired. Let us hope that as they gradually reap the benefits of the trickle-down economy, they turn to the Scandinavian countries as models of sustainable development rather than look anywhere else – including to China.

I have said little about the future of India. That is deliberate. A country that still operates on the basis of a caste system has little chance of serious development for any except those born into the middle class. India will remain a desperate case while the caste system continues to exist.

THE SCANDINAVIAN MODEL

In Chapter 9, we focus on why the Scandinavian countries are different and, before that, how they are different. They are lauded for their environmental values and attitudes that underpin their national and international policies. They are praised for their willingness to give foreign aid freely and much more of it than other rich countries, for their eagerness to be peacemakers and peacekeepers, for their gender equity, for their concern for the young, and for the money they spend on education. If this is not enough, they have high social capital – they are great places to live. All this praise is justified.

Yet under 1000 years ago, the ancestors of modern Scandinavians were the wildest rapists and pillagers and the most feared raiders of Europe and beyond as they extended their forays from nearby targets into the distant Middle East. More accurately, some were the feared Vikings but not all were men of the sword. Once they conquered a new land or territory, they set about farming and developing laws by which the new countries could live productively and peacefully. By and large most Vikings were free farmers with a strong interest in establishing democratic governments. The few famous Viking explorers and raiders – the latter truly ruthless – gave the rest a bad name. We still don't know all the details of how the transformation from Viking to modern Scandinavian occurred. Still, we know about the Nordic enlightenment and the idea that peasant farmers were the social class (and economic base) best suited to developing a progressive society. You and I might challenge this proposition – free farmers as a progressive people is not the conventional wisdom. When and where else have peasant farmers been politically progressive? Nowhere. Never. Well let us dismiss conventional wisdom in the case of the Scandinavians.

The Nordic enlightenment was built on the ideas and initiatives of so-called 'peasants', but who were actually free farmers. In Scandinavia there was not a significant class of shopkeepers as in England or an entrenched aristocracy as in most other parts of the world. There were only free farmers who were neither dull of mind nor wit as farmers are often perceived to be. In the Norse countries, when respect was given it was to the heroes of the past (Eric the Red) or to mythical gods such as Odin and Thor. Respect was not given to an existing higher class. It is possible that no other societies developed on this basis.

There is much in the Scandinavian story to help us towards sustainability. Yet the 'new world' societies of the USA, Canada, Australia and New Zealand have their specific stories. We conclude our case studies in Chapter 10 with the Australian environmental story.

1   Althing literally means 'all thing'.

1

WHISKEY IS FOR DRINKING: WATER IS FOR FIGHTING OVER

Water, water, every where,
Nor any drop to drink.

Samuel Taylor Coleridge, 'The Rime of the Ancient Mariner'

To the thirsty I will give water without price
from the fountain.

Revelation 21:6

Humans have had a fascination with water from time immemorial, yet that fascination has not translated into care for and wise use of this most precious resource – but more on this in a moment. Our interest in water ranges from the religious to the rational. For example, there is the scientific view that life as we know it crawled out of a messy liquid soup in primordial times. One of the results was us – human beings. We are – if you wish to think of it in these terms – about 70 per cent water. Our brains are about 77 per cent water. Coming out of water, we are still a lot of water. The religious view of water runs parallel with the rational scientific view. Various ancient cultures had creation myths about water: stories about how the world began and how we came into existence. A few brief examples are contained in Box 1.

Box 1:
From water we came

In what Nancy Hathaway describes as 'the grandest, most abstract stories, the universe emerges from … an endless ocean (amniotic fluid to some Freudian commentators) …'[1]

For example, there is a Babylonian myth that 'in the beginning, there was only water. The fresh, sweet water was called Apsu … the salt water … was Tiamot, the water goddess'.[2] There is an Icelandic story whereby before the creation of heaven and earth, blankets of cloud hovered above fields of ice in Niflheim. Then we have the Huron Indians, a native American people, who believed that in the beginning there was only water and sky. A Japanese creation myth relates how the world was 'like floating oil, drifting like a jelly-fish in a sea of nothing'.[3]

Go to Egypt where all creation myths began with a 'watery chaos at Nun'[4] or to Greece where creation was believed to have begun when 'Eurynome, the goddess of all things, arose from Chaos (and) separated the sea from the sky'[5] or when – take your pick with the Greek myths – the Greek god Prometheus mixed rainwater (or tears) with earth to make human beings.

Consider *the* flood. We have all probably heard of Noah, his ark and *the* flood. What you might not know is that similar stories are told all over the world, in all sorts of religions and cultures. The idea of the flood could have been exported worldwide by people who lived 5000 years ago on the flood-prone banks of the Tigris and Euphrates rivers (of which we have heard much in recent years). There was a great flood then that devastated Mesopotamia (Iraq).

Thor Heyerdahl describes how children are fascinated with 'the wondrous fairy tale of Noah's ark riding the flood with all the animal species of the world, while adults wonder how he found them all…'[6] He argues that the story is a legend. In a Sumerian text, the ark lands in modern Bahrain, and not on Mt Ararat. The text suggests the flood wave was nearly 8 metres. Consider that the 2004 tsunami that devastated Aceh in Indonesia, Phuket in Thailand and parts of Sri Lanka reached twice that height. Noah's flood could have happened as a result of a relatively small tsunami – if there is a scientific explanation, this might be it.

An alternative theory about the spread of the flood stories is that they were created and then spread after the great thaw that followed the last ice age around 13 000 BC. Any coastal community could be excused for inventing a story of a supernatural force on a mission to teach or torment humans to explain the worldwide rise of sea levels.

Maybe water just resonates deeply in the human psyche. Hathaway suggests the ubiquity of flood stories is connected to the psychological symbolism of water.

ONCE WE WERE SMART

Enough ancient myth – next a little fascinating history. John McNeil in his environmental history writes: 'For most of human history we needed water only to drink. But in the last few thousand years we have relied on it to irrigate crops, carry off our wastes, wash our bodies and our possessions, and more recently to power our mills and machines.'[7] One of the earliest problems of human civilisation was keeping faeces out of drinking water. Towns and cities were sited near rivers, springs or lakes for the obvious reasons that they provided fresh water for drinking, washing and watering gardens. As human numbers increased and concentrated in urban settings, water from outhouses, cesspools and animal stables seeped into groundwater, while runoff polluted the surface water. There are poor, desperate parts of the world where this is still the situation today. Lack of clean water or sanitation kills some 1.7 million people each year, 90 per cent of whom are children.

What is frustrating is that we were smart enough in the past to solve the technical problems of fresh-water supply and sewage disposal and that we have been using irrigation systems for 9000 years. By 2500 BC, the Sumer and Indus valleys had aqueducts that brought water from the mountains to supply public baths and fountains. Sewers took wastewater out of the cities then put it back into the waterways on their way to open seas or oceans where dilution was the solution. Sumerian homes had water-flushing latrines connected to this type of sewer system.

Diogenes takes a bath

My favourite water story from ancient times is this account of Diogenes: 'Diogenes enjoyed going to the public baths and would spend many a blissful hour there. On one occasion, finding the water dirty, he asked the attendant, "Tell me, my good man, where do they wash who bathe here?"'[8]

Take another example: Roman aqueducts and sewers were one of the marvels of the ancient world. Water was distributed throughout the city by lead pipes and hollowed-out logs. It went to public baths, fountains, 144 public toilets – some with 40 seats – and shops and private homes. A visitor to Rome today will see remains of these wonderful engineering and environmental feats. Alice Outwater suggests that water use for each person in ancient Rome was 1360 litres per day, which is around three times what it is in the USA today.[9]

Unwashed for 1000 years

Outwater also claims that Europeans went unwashed for a thousand years. Why? With the ascendancy of the Christian church, the wealth that the Romans had invested in aqueducts and sewers was spent instead on cathedrals. As she states, 'the Church was concerned with loftier matters than its flocks' temporal existence'.[10] So this ancient yet excellent technology was to be lost for millennia and only became commonplace again in the last 100 or so years!

Today, the same type of choice is faced in the grossly overpopulated cities of the world, the difference being that the church has been replaced by the golf course. If we are to accelerate sustainable development in Third World cities, should we spend money on five star hotels and golf courses for the rich and wait for the trickle-down effects to generate the financial resources to put in sewerage in the cities, or should we make the provision of sewers and clean water first priority, and let other developments follow? Of course, the well-off can afford to pay the hotel tariffs and golf club charges. But it is not obvious where the money is to come from to provide public water and sewerage infrastructure – other than the World Bank.

Fascinating as it is, we don't have the space in this book to more fully explore the history of water and wastewater. We leave the ancient world and take up the story with the coming of the Industrial Revolution a couple of hundred years ago. Cities grew rapidly in the United Kingdom, then Europe, the USA and elsewhere later. There was no proper sanitation and cesspools were the norm. The dead were buried within city limits. Garbage was dumped in the streets to be eaten by bands of feral pigs. Chamber-pots were emptied out of bedroom windows. Our great-great-grandparents would have done this. The first bathroom in the USA to have a bathtub, sink and toilet was built less than 200 years ago – in 1810. Horsepower relied on horses and manure was as ubiquitous in cities as bitumen and cement today. Dead animals littered the streets. All this waste was washed or shovelled into the nearest stream or lake before the advent of modern sanitation. Towns stank. The taste of water was terrible. People, many people, became very ill and died young.

Sanitation engineering saves the West

Attributes such as clean streets and water were not considered that important in an era when no-one washed for long periods and everyone drank beer, gin, whiskey or rum. What started to change attitudes and lead to the revolution in sanitation engineering was disease. Smart people made the link between polluted water and human ill-health. There was probably no more important environmental reform than the revolution in sanitation. The people who led the campaigns for clean, healthier cities – the environmental activists of yesteryear – were mainly groups of health-conscious women.[11]

The dramatic turnaround in public sanitation and water provision in developed countries (about 100 years ago) proves that if an issue gets the attention of an educated middle class, anything is possible and in a short period of time – as long as the technology is available or is not difficult to develop.

Cost was not a significant issue – even though it was extremely high – when the industrialised countries set about installing sanitation and providing safe drinking water. The major reason that the high cost was not a political hurdle was that governments could recoup it over a long period of time via land taxes (called 'rates' in some countries). Governments borrowed the money to build the infrastructure. Recipients of the services (if landowners) paid a small amount each year until the capital and interest were repaid. Because the upfront borrowing was done by governments in countries where sovereign risk was low, it was easy to obtain the money. In contrast to this, the poor developing countries do not have, in the main, an efficiently administered pricing system for sewerage and water infrastructure and consequently sovereign risk is high, which means they cannot finance the much-needed services. However, it is not beyond their political and administrative capabilities to change this. They must if they are to develop.

DRINKING THE LABEL

Today our fascination with water is shown in its display as a fashion statement or minor positional good. To be in the latter category it needs to be obviously foreign and high-priced. On planes and trains, at football and tennis matches, on the beach and out beyond, we like to be seen with a bottle of designer spring water. Think of the upmarket brand names (Evian,

Whistler) for the important occasions and the run-of-the-mill names (Frantelle, Cool Ridge) for everyday consumption. The more exotic the source, the more prestigious the water. We pay good money for this stuff. It costs much more than petrol.[12] Bottled water in Thailand with enticing names such as Aura, Siam, Namthip and Mount Fleur sells for 50 to 60 cents Australian per litre. A bulk purchase of local brand Sprinkle – you've got to like the name – comes at 30 cents per litre. At least the Thais are drinking the local water! I often wonder what Archimedes would have thought about Australians drinking bottled water shipped from Fiji or further afield from Canada or France. The displacement of water by water!

Notwithstanding our willingness to pay a high price for fashionable bottled water, we take a lot of convincing that we should pay for water out of the tap – more precisely, that we should pay according to the amount we use. The convention applied in most developed countries, by which a minimal but sufficient allocation is 'free'[13] before a pay per litre charge sets in, is a model that meets both equity and efficiency criteria. To date, water managers are guessing as to the proper price for water – we don't know enough about the value of environmental flows, the recharge of groundwater or the price responsiveness of demand (by both urban dwellers and farmers). Still, a guess at the appropriate price is better than not charging at all.

Fill the bottle from the top

I did some calculations recently on the average cost of water in my city of Brisbane, Australia. It was 82 cents Australian per 1000 litres from the tap, compared to up to $4.00 per litre from the bottle. The bottle does cost something, but the difference is absolutely fantastic. We might use tap water to hose down concrete driveways and footpaths, but the bottled water we will be careful not to spill. As I said, we are fascinated by water but are yet to be serious concerned with its wise use. For a few facts on water see Box 2.

> **Box 2:**
> **Water**[14]
>
> - 2.4 billion people lack access to improved sanitation.
> - 1.4 million people lack access to clean water.
> - A 10-minute shower uses 45 litres of water.
> - The average bath uses over 181 litres of water.
> - A standard toilet flush uses over 9 litres of water.
> - 50 glasses of water are required to produce one glass of orange juice.
> - About half a kilogram of beef requires over 11 000 litres of water to be produced.
> - A Thai village of 60 000 people uses 6500 cubic litres of water per day. One golf course in Thailand uses 6500 cubic litres per day.
> - The cost of a bottle of water is up to 1000 times the cost of water from the tap.

ABUNDANT BUT SCARCE

Fresh water is abundant on a global scale but is far too often scarce where it is needed. Figure 1 illustrates the fact that of all the water that exists in the hydrosphere, only the tiniest proportion is available as fresh water for immediate human use. There is in the order of 1.4 billion cubic kilometres of water in the hydrosphere; of that only 3 per cent is fresh water and most of it is in icecaps and glaciers, leaving precious little that we can get our hands on relatively easily.

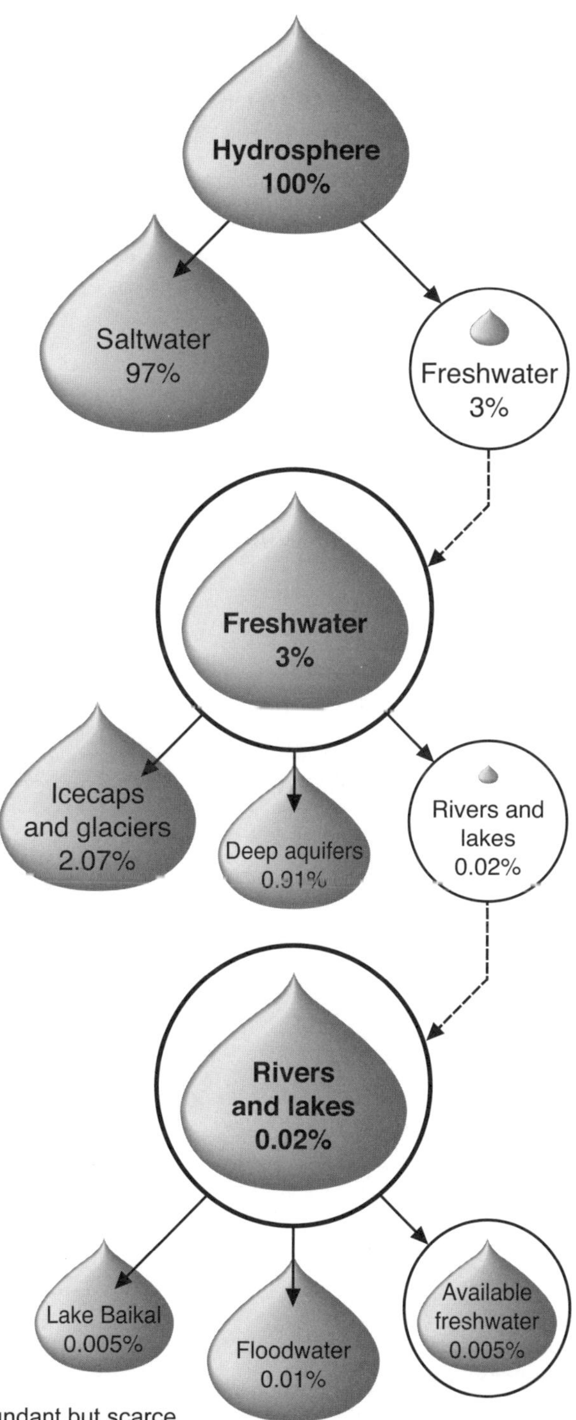

Figure 1: Abundant but scarce

The world's expected population increase of three billion people by 2050 (at which stage the global population might stabilise) will put enormous pressure on fresh-water supplies. The relatively minor skirmishes we have witnessed over sharing water will be nothing like those to come. Box 3 lists some water-related conflicts between nations. The only continent to miss out – and so it should as a single-nation continent – is Australia. There the conflict is between communities and states (provinces).

Box 3: Interstate water-related conflicts[15]

SHARED WATER RESOURCES	COUNTRIES INVOLVED IN CONFLICT	CAUSES OF CONFLICT
Amu Dar'ya and Syr Dar'ya	Uzbekistan, Kyrgyzstan and Kazakhstan	water scarcity and control
Danube	Hungary and Slovakia	control (diversion) of water
Ganges	India and Bangladesh	water scarcity and diversion
Great Lakes	USA and Canada	water pollution
Han	North and South Korea	control of water
Indus	India and Pakistan	water scarcity
Lake Chad	Nigeria, Chad and Cameroon	water sharing
Lake Victoria	Uganda, Kenya and Tanzania	water sharing
Lauca	Bolivia and Chile	control of water
Mekong	China, Burma, Thailand, Laos, Cambodia and Vietnam	water scarcity and control
Orange	Botswana, Lesotho, Namibia and South Africa	water scarcity and diversion
Paraná	Argentina, Brazil and Paraguay	control of water
Rhine	France, Germany and the Netherlands	water pollution

Rio Grande	USA and Mexico	water pollution
Sahan aquifer	Libya, Egypt and Sudan	water scarcity
Salwan/NuJiang	Burma and China	water scarcity
Senegal	Senegal and Mauritania	water scarcity
Szamos	Hungary and Romania	water pollution
Okavango	Namibia, Botswana, and Angola	water scarcity
Zambezi	Zambia, Zimbabwe, Mozambique and South Africa	water scarcity and diversion

Consider that, in 1995, 92 per cent of the world's population had 'relative sufficiency' of water, but by 2050 only 58 per cent will have this.[16] It is forecasted that in 2050, 18 per cent of the world's population will face water scarcity and 25 per cent water stress, where scarcity is defined as a situation whereby under 1000 cubic litres is available per person, which according to Pearson means that 'chronic water shortages impede economic development and cause environmental degradation'.[17] We are going to have to find means to save water, and to share water. Both are possible.

SOLVING THE WATER PROBLEM

Various authors have turned their attention to solving the water problem, but here we focus on three contributors to Bjørn Lomborg's 2004 book, *Global Crises, Global Solutions*. Frank Rijsberman supports low-cost (community-managed) water supply and sanitation for the world's poorest communities. He also suggests reusing wastewater for peri-urban agriculture, encourages smallholder agriculture in wetlands (as opposed to draining wetlands to become regular arable land), and supports research and development to increase the productivity of water in food production ('increasing the crop per drop'[18] – as he says). John Boland writes that 'water is different' to market goods and is a matter that should not be dismissed out of hand. That water is essential for life has led many to argue that it should not be allocated on a basis of what one is willing and able to pay. Some people starve, some are malnourished and some are obese because food – another essential ingredient of life – is distributed according to the market mechanism. The same rich–poor dichotomy would happen if the means to obtain water was the ability to pay for it. Boland argues that water 'in the environment' (that is, 'everywhere around us') is a public good in the way economists define public goods; it would be, he claims, 'impractical or impossible to levy a charge for water in the environment, and then exclude non-payers'.[19] He discusses water that is collected, stored and reticulated to be used in cities and on farms when it is needed. Because what is supplied to one person cannot be supplied to another person, water so managed is a private good. But who has the right to convert water from a public good to a private good by stopping its flow in a river or collecting it in a tank? This is the most basic – and fundamental – question about how humans use water and there is no agreement on a universal rule – based on ethics, ecology and economics.

Finally, water has to be safe to use and there has to be a safe means of disposal of wastewater. This is where enormous externalities occur. Consider the benefits (both monetary and in human wellbeing) if there was a significant reduction in illnesses and death by eliminating waterborne diseases (see Box 4 and Table 1). The benefits from safe supply and safe disposal of water do not go to just the user but to the community as a whole. Here again water takes on a characteristic of a public good. These three situations are useful pointers to public policy on water use. Water is not simply water, we have discovered, but we are still to determine rules that meet sustainability criteria for allocation.

Box 4:
The loss of life and health through water-related diseases

In 1993, the Harvard School of Public Health in collaboration with the World Health Organization (WHO) and the World Bank began a new assessment of the global burden of disease (GBD). This effort introduced a new indicator – the disability adjusted life year (DALY) – to quantify the burden of disease. The DALY is a measure of population health that combines, in a single indicator, years of life lost from premature death and years of life lived with disabilities.

Table 1: Deaths and DALYs from selected water-related diseases, 2000[20]

DISEASE	DEATHS	DALYs
Diarrhoeal diseases	2 019 585	63 345 722
Childhood-cluster diseases		
Poliomyelitis	1136	188 543
Diphtheria	5527	187 838
Tropical-cluster diseases		
Trypanosomiasis	49 129	1 570 242
Schistosomiasis	15 335	1 711 522
Trachoma	72	3 892 326
Intestinal nematode infections		
Ascariasis	4929	1 204 384
Trichuriasis	2393	1 661 689
Hookworm disease	3477	1 785 539
Other intestinal infections	1692	53 222
TOTAL	2 103 275	75 601 027

Note: One DALY can be thought of as one lost year of 'healthy' life. This table excludes mortality and DALYs associated with water-related insect vectors, such as malaria, onchocerciasis and dengue fever.

One thing that Henry Vaux, the third contributor to Lomborg's book, does that the previous writers don't is focus on groundwater. This type of water accounts for about one-third of the world's usable water supply. He makes the point that 'ground water overdraft, in which the quantities of water pumped exceed the quantities recharged, threatens the sustainability of many of the world's ground water sources'.[21] He furthermore draws attention to the fact that to feed the world's growing population more irrigated agriculture is going to be required, and 'water shortages will soon manifest themselves as food shortages'.[22] Reverend Thomas Malthus, doomsday prophet of the eighteenth century, who said we could not feed an ever-expanding human population, could yet be proved right! Vaux calls for 'reallocative institutions which facilitate the movement of water from relatively low-valued users to higher-valued users'.[23] This is commendable as far as it goes. Having water users – whether in industry or agriculture or domestic users – compete via price is the economists' solution.

However, if every drop, including the very first drop, was allocated solely by price, the poor would go without – and die. There must be a global charter, or an agreed ethical principle, that states that basic needs must be met – at no cost to the user – before water is put on the market. If key human and environmental needs are first met through a regulatory cap being set, the market solution is the best allocation method thought of so far to allocate the remainder. Let us consider why – and any problems.

Worldwide, water is wasted on low-value farming. The monetary returns per unit of water applied to a hectare of land should be the yardstick. Very sensible! Yet this approach will favour the wealthier consumers of agricultural products and farmers who are able to meet the former's tastes.[24] Is it not likely that wine grapes will be watered before potatoes? What we witness today – with the rather ridiculous prices paid for bottled water in the rich countries while slum children drink rat-infested chemical-polluted water flowing through their villages – is not what we need in agriculture. This is not to argue against a well-thought-out user-pays system for irrigation water. Obviously we have not solved the water allocation problem yet. The root cause is the existing unequal distribution of wealth and income in the world.

Water is different

Water is different. At least until the basic needs of all are met – and all externalities measured and taken into account – it should not be allocated by the price mechanism. Acceptance of this dictate would not prohibit 'pay by use' charges in the industrialised countries. However, allocation between countries, and within developing countries, has to be based on an equity rule (a 'golden rule' of treating others as we would wish to be treated). Within countries this policy should be easy to achieve if there is a strong government committed to this goal. Allocation between countries is going to require an international law far more difficult to negotiate than the United Nations Law of the Sea, which took decades of talk and negotiation before it was put in place – and today is still not recognised by the USA. To achieve a trans-boundary water regime – something much more difficult – should be considered one of the global community's major challenges but not an impossibility. Let us take on such challenges. We put a man on the moon. We make messages such as the words in this book shoot effortlessly at fantastic speed across the globe. If we want to we could easily solve the water allocation problem on the basis of fairness.

We would have far less trans-boundary conflict if the borders of nation-states were based on catchments. The people of a nation-state are inclined to heed the plight of their less fortunate citizens. But let us not indulge in utopian dreaming.

Water allocation and use has grabbed the attention of scientists, environmentalists and politicians, if not the average citizen. Here is an example of an occasion when the big issues surfaced. In March 2003, over 24 000 people attended the Third World Water Forum held in the Kinki region, Japan. The fundamental debate was 'whether water should be considered a commodity to be bought and sold in the market. This dichotomy we have already noted. A subsidiary debate was whether community-scale projects were preferable to large-scale dams. Another was over the role of the private sector in owning and operating water infrastructure. And yet another over the Candessus Report, a study commissioned for the conference that dealt with the financial means of achieving the UN's Millennium Development Goals. Michel Candessus is the former managing director of the International Monetary Fund (IMF). One can appreciate in a meeting of 24 000 people he would face antagonists. For what must have been an enormous amount of talking in Japan, there was little to take us forward from the meeting. It will be a sure sign of political cowardice if, in 20 years time, we engage in yet another talkfest on water in which we are still searching for solutions.

EQUITY AND EFFICIENCY

I have held the view for some time (about 25 years) that the poorer cities of the world urgently need the sanitation revolution that the Western cities experienced 100 or more years ago. The economic benefits in terms of reduced health costs, if nothing else, would be enormous. I do not believe that 'small is beautiful' in large-sized towns or cities, hence citywide sanitation and a water supply are needed. However, the methods of providing clean water – large dams, desalinisation, rainwater tanks, groundwater extraction, or some mix of these – will vary from place to place. The more important issue is to get the right mix of equity and economic efficiency in any water policy.

I believe we have the chance to do this with water; something we don't have with that other essential necessity for life – food. My reason is that we have not yet formed an attitude that water is, or should be, privately owned – and therefore the idea of sharing is not out of bounds. Except for in a few remaining tribal societies, food is not freely shared. It is a private good. While, as argued above, water needs to be allocated to its highest value use (a litre applied to an edible crop is more valuable than a litre washing a city street), most societies recognise the fundamental principle that one does not deny a thirsty person a drink. This principle overrides all else if you are a humanist. This is no different to the proposition that the preservation of human life (the prohibition of murder) overrides the exercise of power (by criminals or dictatorial rulers) to take the material possessions of others by killing them. Not providing sufficient water for life is a criminal act.

1   Nancy Hathaway, *The Friendly Guide to Mythology*, Viking: New York, 2001, p. 1.

2   Hathaway 2001, pp. 3–4.

3   Hathaway 2001, pp. 13–15.

4   Hathaway 2001, p. 18.

5   Hathaway 2001, p. 23.

6   Thor Heyerdahl, *Kon Tiki*, 2005, pp. 30–1.

7   John McNeil, *Something New Under the Sun: An Environmental History of the Twentieth Century*, WW Norton: London, 2000, p. 118.

8   George Pavlu, *Diogenes: An Anecdotal Biography of the World's Greatest Cynic*, Methuen: North Ryde, NSW, 1987, p. 54.

9   Alice Outwater, *Water: A Natural History*, Basic Books: New York, 1996.

10   Outwater 1996, p. 135.

11   Drew Hutton and Libby Connors, in their *History of the Australian Environmental Movement* (Cambridge University Press: Melbourne, 1999), tell the story of various women's organisation at the beginning of the twentieth century.

12   Of course, you can refill your Evian bottle from the kitchen tap for a cent or two.

13   By 'free' I mean the storage and reticulation costs are covered in a general tax (or rate) on land or dwellings.

14   These statistics have been adapted from Anita Roddick and Brooke Shelby Biggs, *Troubled Water: Saints, Sinners, Truths and Lies about the Global Water Crisis*, Anita Roddick Books: West Sussex, 2004.

15   This table has been reproduced from Ron Nielsen, *The Little Green Handbook: A Guide to Critical Global Trends*, Scribe: Melbourne, 2005, p. 273.

16   Ian Pearson (ed.), *The Macmillian Atlas of the Future*, Macmillian, New York, 1998, p. 47.

17   Pearson 1998, p. 47.

18   Frank Rijsberman, Sanitation and access to clean water, in Bjørn Lomborg (ed.), *Global Crises, Global Solutions*, Cambridge University Press: Cambridge, UK, 2004, p. 509.

19   John J Boland, Alternative perspectives, in Lomborg 2004, pp. 530–1.

20   Adapted from World Health Organization, *World Health Report 2001 – Mental Health: New Understanding, New Hope,* Version 2 data tables on the global burden of disease, Geneva, 2001, http://who.int/whr2001/2001/.

21   Henry Vaux, Perspective paper 9.2, in Lomborg 2004, p. 537.

22   Vaux, in Lomborg 2004, p. 538.

23   Vaux, in Lomborg 2004, p. 540.

24   Because the much higher number of urban dwellers have the financial ability to outbid farmers, politicians (with an eye to rural voters) are unlikely to favour a genuine, market-based water allocation system. An example of this was a newspaper report 'Water transfers "won't get PM's nod"' in *The Australian* in June 2005. The prime minister of the time was John Howard.

2

FISHING FOR FOOD WITHOUT THE HELP OF MIRACLES

Life is saltfish.

Halldór Laxness, Icelandic Nobel Laureate for Literature, 1955

Food and fresh water are the most basic of requirements for the world's people. We produce enough food to feed the present world population at a reasonable standard. The problem is the lack of a fair distribution system. At this point in human history you can eat as much as you like – and what you like, including the most costly meal in the world – if you have the money to pay for it. Here we will focus on a somewhat different food issue, although it is also about a conflict between the powerful and the weak.

FEEDING THE MULTITUDE

Humans have been fishing for food (fish are a major source of animal protein) from ancient times. Christians know the story in the Bible about the feeding of the multitude with fish.

Most of us are probably unaware of an early piece of science fiction that elevated fishing to one of the two planks of the modern economy. Writing in 1733, the Irish satirist Samuel Madden published *Memories of the Twentieth Century*, in which he predicted that two giant companies would dominate the world in what then was a distant future time. One company was the Royal Fishery, the other the Plantation Co.[1] I note this for the sole reason of highlighting the importance previous generations placed on seafood in their diets. There was no royal dairy or royal abattoir. What was the case in the past is so today and probably will also be the case tomorrow.

More people, no more fish

Until the rapid population growth of the past century there was little concern that our numbers would threaten to deplete fisheries' resources. The human population exploded after the Industrial Revolution. It doubled between 1960 and 2000, and by 2040 to 2050, or thereabouts, it will be nine billion. World population growth is putting extreme pressure on the world's fisheries. Given that we are today at the limit of our wild-catch fisheries resources, what do we do to guarantee sustainable fish harvests with a 50 per cent increase in human numbers in the near future? To make matters worse for the status of fish stocks, the affluent middle class are turning to a healthier Mediterranean diet based on seafood. Good for them, but what of the fish stocks?

Today, of the world's six plus billion humans, one billion rely on seafood as their main source of animal protein. The amount eaten varies significantly from country to country (in fact, from locality to locality), according to the availability of seafood, its cost and local culinary traditions. Population growth rates also vary significantly around the world, as does productivity and income. This means there is no one prediction we can make about the effects of the increased demand for seafood, or our capability of meeting it. However, the general trend is quite evident.

INCREASING, THEN DIMINISHING RETURNS

Through the 1950s and 1960s to about 1970, the world fish catch was increasing by about 6 per cent each year. This is simple to explain. Imagine fishing offshore during the period 1939 to 1945 as World War II raged! Very, very little fishing occurred. It was more than likely that Norwegian and Scottish fishing boats were transporting resistance fighters between countries than seeking food in the seas. After World War II, seafaring was once again safe and vessel technology (based on advancements in naval shipping) had improved markedly. This new technology allowed vast areas of ocean to be explored and new distant fisheries were soon discovered.

At first, average catches increased, then the rate of increase started to decline, to about 2 per cent per annum through the 1970s and 1980s. In the 1990s the world catch levelled off at over 90 million tonnes, and this is where it looks like remaining indefinitely – unless collapses in catches occur. More than 30 million tonnes of seafood now comes from aquaculture and its contribution to the total supplies is increasing. The total available seafood production is approaching 130 million tonnes. Before we get excited about aquaculture solving our seafood demand, we need to be aware that there are limits to what aquaculture can produce, unless there are technological breakthroughs which we have not yet anticipated.

Before we explore anything else, let us put this story into perspective. Figure 2 presents an extremely important insight. It shows that, over the past 500 years, fisheries' production has kept pace with human population growth, proving wrong Malthus's prediction that population growth would outstep food production during this period. But now that we have reached the peak catch of wild fish, the still increasing number of humans will have to rely on dramatically increased output from aquaculture if seafood is to remain a significant part of people's diets – and the latter is what people worldwide are expecting.

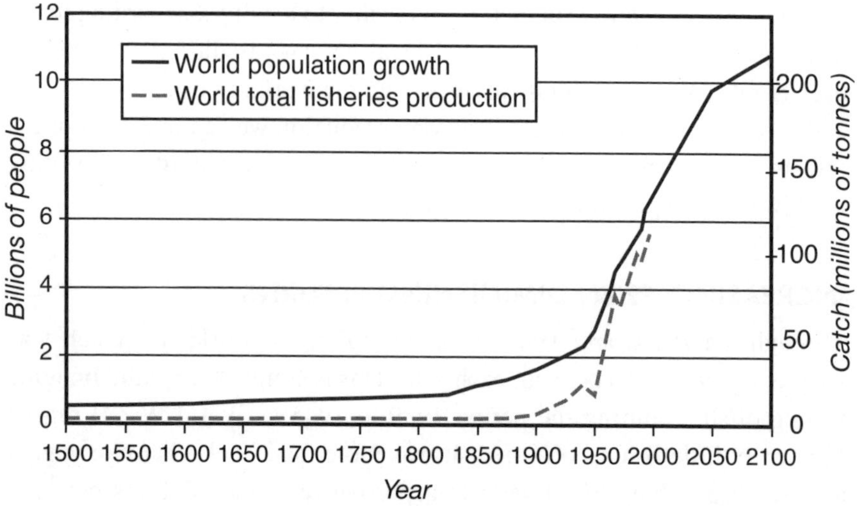

Figure 2: Human population growth and world fisheries' production[2]

Aquaculture as a contribution and a problem

At present about 90 per cent of aquaculture takes place in low-income developing countries. With ever-increasing worldwide demand for seafood, this will be a major economic plus for these countries. Export income should increase and this will induce expanded production. Any poor country with the appropriate environmental conditions to build aquaculture ponds should be a serious player in aquaculture production and some countries with poor environmental standard policies will benefit – at least while they get away with the pollution.

How aquaculture is undertaken, in what circumstances and with what levels of environmental protection are fundamental issues to be resolved. In the past there have been serious problems with shrimp aquaculture in South-East Asia. Mangroves have been destroyed, so was rice-paddy land, and vast areas of shrimp farm ponds became unproductive within a few years. To add to the externalities, local waters were polluted. As a consequence of unpriced environmental costs, shrimps (called prawns in some countries) are put on international markets at a half, and sometimes a third, of the price of First World shrimps. If cost is all that matters to the consumer, the end is in sight for shrimp production in the First World, with its extremely high environmental standards.

RECALL THE FISH WARS

Before pursuing the potentially disastrous future of aquaculture there is a short story from the recent past that provides perspective. How easy it is to enter into conflict over natural resources – wars are often fought over nature's gifts. If one doubts that nations won't go to war over environmental resources, think cod. This is the fish that was in the title of a book, *Cod: A Biography of the Fish that Changed the World,* by Mark Kurlansky, which became a bestseller. The 'cod wars' started not long after Iceland's fisheries were constrained by a bilateral law, enacted by the Anglo-Danish Convention of 1901, that the Icelandic people did not write or agree to, which put a three-mile limit around Iceland's shores.

Iceland, a free Scandinavian country (the Icelandic people had come from Norway originally – some via Scotland and Ireland – more than 1000 years ago), was for some time during the fairly recent past a Danish 'colony'. It was during this period that laws and international treaties disadvantaging the Icelandic people were put in place. When Iceland became a nation free to exercise its sovereign powers in 1950, it extended its territory to four miles offshore. Fishing was its mainstay.

Small is beautiful

Iceland has the propensity to play in the big league, rather than the trivial one that most would assign to a nation of 300 000 people living in the frozen backblocks of the Arctic. Other then a temperamental singer with the one-name, Bjørk, what does the rest of the world know or care about Iceland? Yet, today, it is number one in the United Nations ranking of all countries. In late 2008, Iceland got caught up in the so-called 'derivatives and leverage' bubble that had its genesis in deregulated finance. The International Monetary Fund (IMF) came to Iceland's rescue. It is ironic, given the cod wars between Iceland and Britain, that the latter was subject to an IMF rescue in 1976 – Iceland should not feel bad about its situation given this fact.

Following the leader

We should note that in a past era (the 1950s) any offshore territorial claim was, to quote Mark Kurlansky, 'highly questionable'.[3] The USA started the ball rolling in the other direction in 1945 when President Harry Truman proclaimed that the USA had the rights to mineral resources (including oil) on the country's continental shelf. From the American perspective there was to be no three-mile sovereign limit, no 12-mile limit, nothing! The powerful could take control of their whole continental shelf. The American proclamation stirred numerous nations into responding by claiming their own continental shelves. The British and Europeans, in general, did not support the notion of extended sovereignty. Hence, the Icelandic claim (more aligned with the American position) was not acceptable to the British, who were the main competitors for the cod.

Are cod worth fighting for?

In the period we are considering, the Icelandic fishers produced an income that went a long way to support the whole country. They were never going to be satisfied with a four-mile territorial sea. In 1958, Iceland extended its territorial limit to 12 miles and the first cod war began. Let us happily note that in the three cod wars not once was war declared and no-one died. Of this war, Kurlansky writes that France, Belgium, Denmark, Germany, the Netherlands and Spain backed the Brits. This was 'as close as Western Europe has ever come to a united front'[4] – at least before the European Union came into existence. Against tiny Iceland much of Europe had united. There was no recognition of the fact that, as 'the Greeks of northern Europe', the Icelandic people had given Europe its first parliament, over 1000 years ago. The Icelandic people could claim to be the progressive people of Europe, yet they were being hassled, hindered and harassed by the modern European powers.

David and Goliath

As consequence of the 'outlandish' Icelandic initiative, as the British described it, the British fishing vessels retreated from the 12-mile limit around Iceland to fish further out to sea, by then accompanied by British warships. A minimum of 37 Royal Navy ships were involved at any point in time. Repeat that number – 37! This amounted to more than one British ship for every 6500 Icelanders. The Icelandic people must have been fearful giants of Nordic mythology (Vikings, if you wish) to warrant such attention! On the opposing side in this war was the Icelandic Coast Guard consisting of seven vessels, each with one gun and manned by policemen or civilians. No navy, no army. These non-military people were good seamen (seamanship is an Icelandic tradition, as Eric the Red proved when he discovered North America 500 years before Columbus) but had no fighting background – leaving aside their Viking background of 1000 years before. The Icelandic seafarers managed to arrest one British ship in the two-and-a-half years of the first cod war. A temporary solution and end to the war came in 1961 when Britain recognised the 12-mile limit: Iceland one, the UK zero.

Slow learners

By March 1971 they were at it again. Iceland declared its boundary to be 50 miles out to sea. There was no reason in their eyes as a fishing and seafaring nation to do otherwise. Had they not discovered Greenland – and North America? If it was good enough for the USA to claim seawards to the extent of its continental shelf, Iceland felt free to make a less controversial claim. Britain and Germany took Iceland to the newly established International Court of Justice, but Iceland refused to recognise the court's jurisdiction, arguing that the continental shelf belonged to the adjacent nation. There was to be no turning back on what the USA had started in 1945.

Precedents are precedents, and those who start the ball rolling are always in danger of being overtaken and pushed aside by those who do nothing more radical than follow suit. The Icelandic people were experts at pushing out the boundaries. They had been doing that since the first Norwegian Vikings settled there – after Iceland there was Greenland then North America. What was a mere 50 miles in the 1970s?

The British will never forgo their fish and chips

The second cod war was shorter than the first but the firepower of the Icelandic Coast Guard had improved and casualties were a new distinct possibility – except that the Icelanders had no desire to hurt the Brits. Rather than firing bullets (although a little across-the-bow shelling did occur) they used a new 'weapon', a trawl wire-cutter, a piece of machinery that separated the British fishing net from its boat. In one year, 84 British trawlers lost their nets[5] and the cod war degenerated into 'dodgem cars on the high seas' as Icelandic vessels manoeuvred to cut nets and the British manoeuvred to escape this embarrassment. Had not the British come out victors in World Wars I and II? These were, of course, principled wars, but to fight a nation of 300 000 people over fish and not win was, well, embarrassing at best.

After the shells-across-the-bow incident on 18 March 1973, the British trawlers withdrew to 50 miles offshore. Incremental retreat seemed to be the British strategy. However, there was a limit and at some stage the Icelandic people would win. The war was about to end. Britain eventually accepted the new Icelandic fishing zone. Yet, cod-dependent Iceland was not finished.

In 1973, the UN gave credence to the concept of a 200-mile exclusive fishing zone. The only European countries to back this were Iceland and Norway. On 15 October 1975, Iceland extended its fishing zone to a 200-mile limit. Iceland was starting to look like a powerful state – doing the sort of thing the USA would do and getting away with. The British and German fishing fleets refused to leave the greatly extended Icelandic zone and Cod War III commenced. It was to be short and serious. In a five-month period there were 35 ramming incidents but Iceland won – and not an injury on either side.

It was not the end of the world

The British had gone to war three times with a Viking nation because British fishmongers had convinced the population that it would be the end of the world as they knew it if their favourite meal, fish and chips, was to disappear.

I must point out that cod is not the only fish that could accompany chips. The English were drawing a rather longbow – curry was fast closing in on fish and chips as the new traditional take-home or dine-out meal and would surpass fish and chips by the 1980s. The British search for cheap labour from its ex-colonies (India, Pakistan, the West Indies and numerous African ones) had a dramatic impact on popular food tastes. This is how curry made number one. Thankfully Iceland had won the cod wars. Britain might have been embarrassed by an oversupply of cod as its cuisine changed. That aside, there were other fish species – such as pollock, whiting and redfish – to replace cod and any of these other species tasted as good in fish and chips. The whole affair was seriously silly.

Years after the cod wars had ended, while on sabbatical leave at the University of Manchester in 1983 (of all things writing a book about Australian fisheries), I visited the once-great English cod port of Hull to be reminded of the unfortunate consequences of the cod wars. The large

British fleet of cod boats was permanently tied up at the wharf, where it had been for years, rusting away. Sad!

SO MANY FISH IN THE SEA

One war over fish is enough. Given the expected mismatch between world demand and supply of seafood,[6] what do we do? At a global level we face a major constraint. We have a finite planet, with finite amounts of ocean and land. We ask the globe to feed us, provide shelter and clothing and assimilate waste, with the only input being energy from the sun. There are only so many fish in the sea, and miracles only occur in the Bible. Of course, technological advances can allow us to search ever-deeper waters for fish. But, as always, there is an absolute limit.

Coming to these limits, we can try another tack. By changing our diets we can feed either more (or less) people given the resources we have. Eating grains is the answer if we want to feed more people. Continuing on our self-destructive path of population growth means we will have to feed more, until the day comes when we will, at long last, recognise that Malthus's prediction was right. At a global scale – applying principles of fairness – we have already overshot the limit.

To get people to change traditional diets is not easy. And where we witness changes in diets, they are about eating further up the food chain rather than down the chain, which means more seafood and other meats, displacing vegetables and grains. There has been for some considerable time a worldwide trend to consume more high-quality protein, in particular seafood, meat and soybeans. Nothing points to decreased demand for seafood. We have a problem.

Closing the commons

With regard to global wild-catch harvests, the major objective has to be not to over-harvest the seas. Fish, like other renewable resources, can be harvested to produce a sustainable yield, year in, year out. That needs to be the goal. In the past there was a major institutional hurdle to achieving this. The oceans and seas were treated as a 'commons' to which all had access (called 'open access' in the fisheries literature), the consequence of which was a rush to catch as many fish as possible before someone else did. It makes sense! Countries competed for fish, as did the UK and Iceland in the cod wars.[7] And fishers competed among themselves. This was the only rational thing to do in the situation of open access. Inevitably this has led to over-fishing and, in a number of cases, it has resulted in the depletion of fish stocks and closure of fisheries.[8]

In the days of open access, not only did the fish stocks suffer, but there also was an enormous waste of fuel, labour and capital in the rush to catch fish. Bigger and bigger boats were built with increasingly more powerful engines and more effective nets. Much more fuel was burnt and hence wasted, increasing greenhouse gases. Fewer and fewer fish were harvested per trip, notwithstanding the significant increase in effort. Once the fish were gone, the boats were tied up and rusted away or, worse, were deployed to another fishery in another country, which was consequently destined to the same sad fate.

The tragedy of the commons

The problem of open access for fisheries (Garrett Hardin's 'tragedy of the commons') is now resolved at a global scale, yet there is an exception to which I shall come. The declaration of the 320-kilometre exclusive economic zones (EEZs) by maritime states means that a country has full responsibility for fisheries in its much-extended territory. Iceland won the war for extended territory and, in doing so, changed the world. There are still some remaining 'high seas' beyond any nation's control, and trans-boundary stock (pelagic fish such as tuna) cause international management coordination problems as they move between national zones.

Whales cause their very own problem. Whales, not fish, tend to excite people more than any other creature living in the sea. Still, there has been a fundamental regime change in the governance of the world's oceans. The principle of taking no more than sustainable yields[9] is universally accepted (if not enforced), and management systems based on limited entry (only so many boats and so many nets, lines or whatever) or total allowable catches (TACs) are imposed in many cases. All this sounds good. However, the practice is not necessarily in keeping with sound theory. A major problem is that many developing countries lack enforcement of otherwise good fishing laws. Illegal fishing in both home country seas, as well as foreign waters, continues to be a significant problem. Fish theft is probably the most widespread crime in the world. It may not grab attention in the way the international drug trade does, but it is as universal.

Fish theft

Fish theft tends to be glorified as twenty-first century piracy. In the 1950s and 1960s, Hollywood depicted pirates as heroes; however, is stealing fish in poorly policed waters, such as off the west coast of Africa, something to be glamorised? Local fishermen, generally poor, are harassed by thieves and their nets cut and rendered useless. They catch no fish. Their families starve. The pirates are not the swashbucklers of the seventeenth and eighteenth centuries, but instead are poor, desperate people working for large corporations based in East Asia, from where they sail their boats. There are no heroes in this tragedy.

Those working the virtually unseaworthy pirate vessels are so poor that there is a strong incentive to take the risk of being apprehended and jailed in a foreign country. Maybe jail in a Western country is not such a bad outcome: if fish thieves are caught in a developed country, the time in jail is probably the only well-fed 'holiday' they will experience in their lives. We shouldn't blame the poor fishers, rather the owners of the vessels: rich crooks who ply their trade throughout the poor coastal countries. The vessel owners are never caught – just as drug 'lords' are feted as successful citizens while their couriers face the gallows.

Other fisheries problems

There are many aspects to sound fisheries management. In addition to ensuring sustainable yields of the target species, there are numerous other ecological matters to be dealt with, not least being the impact of fishing on associated species, as predator–prey relationships are affected by fishing. How are predator–prey relationships altered when humans become the top-level predator in the sea and what does this mean to the overall functioning of the marine ecosystem? If humans remove the 'top-of-the-tree' marine predators, because these are the fish we enjoy eating, what happens to all the little fish and other critters that the big fish would have eaten? Then come what economists call the 'knock-on effects' as a new balance has to be arrived at in the ecosystem.

Increasingly, attention is being focused on non-fishing problems such as marine pollution and threats to rare and endangered marine life (such as dolphins, turtles and seals). What this implies is that a whole-of-the-environment approach is now being applied (or at least thought about) in the management of wild-capture fisheries. This approach is long overdue, and will remain difficult because the science involved in understanding the complexity of relationships in the ocean (a place beyond human vision) is difficult.

FINDING ANSWERS TO FEEDING PEOPLE

If the solution to increased demand for seafood is farmed marine life, there is a new set of urgent issues. The major ones are local, not global. This is because the farming of seafood is similar in many respects to intensive agriculture and subject to the same localised environmental concerns. This is the first issue to deal with. The next is that there is increasing interest in price competition between aquaculture production producers in low-cost developing countries and those in high-cost developed countries. This makes for international trade disputes. At present the playing field is not level, because environmental laws are not consistent.

Aquaculture

The history of aquaculture illustrates what should be and what should not be done. In the distant past, fish-farmers (mainly in Asia) had a good understanding of the workings of nature. Their skills were learned over millennia. Growing fish in ponds as an integral part of rice and animal farming has been a part of Asian rural culture for thousands of years and has never caused a problem, environmental or otherwise.

Much of the success of tropical aquaculture has been learned from Asia. Scientists from countries such as Australia went to Asia to learn about fish-farming. (We should note in passing that cold country aquaculture is a completely different story and was developed in Norway and Scotland.) I saw my first example of integrated farming that involved fish culture about 30 years ago in Indonesia. There were quite small ponds (only a few metres square) adjacent to a rice paddy. The farmer had pigs, ducks and fish, all living in a symbiotic relationship. What was waste for one was food for the other. This type of farming is excellent from both an economic and environmental perspective. However, it is not large-scale, and it is hard to believe it ever could be. The quantity of fish produced was sufficient for the family's use and for sale to a local restaurant, but not enough to be a serious business proposition. I ate the farmer's fish at a local restaurant – it was delicious.

What I have described is 'old times' Asian aquaculture. The type of aquaculture that has developed in Asia since has been large-scale shrimp farming whereby hectares of coastal land are excavated and converted into ponds bursting with artificially fed shrimps. This process initially was beset with devastating environmental problems, and still is in some places.

Either good paddy land or mangroves were dug up and converted to ponds. The opportunity cost – that is, the loss of valuable agricultural land and ecosystems – was high. Ponds were side by side in the coastal zone and the polluted wastewater drained into the sea from one pond became the water drawn in for the next pond. Shrimps were swimming in other shrimps' excreta and chemical wastes. Consequently, many shrimp farms failed and a considerable area of paddy land was so degraded that it could not be returned to rice farming. Short-term profits from shrimp sales came at the expense of long-term earnings from rice production. Except for the 'quick buck' (read baht if in Thailand) investor, this was economic and environmental disaster.

Progress proves to be possible

Recently, aquaculture management practices have improved in many parts of Asia and better quality shrimp foods have been developed. From an environmental perspective the future is brighter than it was only ten years ago. However, a vague awareness of the environmental problems in Asian shrimp farming has led to unease in the minds of Western consumers. Asian aquaculturalists have a major task to convince Western consumers that their products are 'clean and green'.

Notwithstanding health and environmental concerns, the cost of Asian shrimps, or prawns, in Western supermarkets (half to a third of that of quality prawns from Australia) is enticing for the price-conscious seafood buyer. Eventually there will come a flood of cheap Asian shrimp and high-cost rich-country farming will go head to head with low-cost Asian farming, while the traditional fisher of shrimps, the wild-catch harvester, will gradually disappear – unless, of course, as consumers we are willing to pay a premium for the genuine hunted-and-gathered prawn. It is in part a matter of taste, a matter of environmental concern, as highlighted in Box 1, and the price in the fish shop.

Box 1:
Fish-finger footprints

Some things are just too silly to make sense. During my research for this chapter, I spent considerable time poking around refrigerated produce displays and shelves in supermarkets to ascertain the extent of 'dolphin safe' labelling, in particular, but I also had an eye out for any other pro-environment or sustainability message on cans and packages. 'Dolphin-free' has caught on, and the occasional Marine Stewardship Council[10] label turned up.

I was not expecting to find packets of quick-frozen Lakes Entrance Australian whiting fillets in with the imported frozen – and cheap – Asian seafood products. Across the supermarket aisle, in the fresh (not frozen) refrigerated display were literally tens of dozens of unfrozen whiting fillets. Why would there be packaged frozen fillets next door, so to speak? Whiting fanciers would not consider the frozen fillets – but maybe at a third of the price, they might. The Lakes Entrance whiting, caught in the beautiful clear and clean waters of southern Australia, had been shipped to Thailand, where they were scaled, filleted and shipped back to, in my case, Brisbane, Australia. This is a return trip of about 15 000 kilometres. I have estimated the saving in labour costs at about Australian 70 cents to $1.50 per kilo. Subtract the shipping costs and the profit is a matter of a few cents. Of course, a few cents per kilo becomes dollars per ton; however, for every 10 ton of fish taking on this round trip, there is one extra ton of carbon dioxide emitted.

There is a very large international trade in seafood, particularly in Asia and Oceania, because of these regions' large populations and dependence on seafood protein. Thailand exports seafood worth US$4 plus billion annually, China is approaching US$4 billion (and growing fast), Indonesia US$1.5 billion, and India and Vietnam both a little less. Australia, which is a minor seafood producer on a world scale – ranked at 50 or lower – exports at least US$1 billion annually.

Not only is an increasing world population driving up the demand for seafood, but so is a change in diet based on the healthy eating movement.[11] And there is the lowering of the cost of seafood products through economies of scale and the use of inexpensive labour in developing countries. All this might seem positive for aquaculturalists and for an exporting country's balance of trade. And so it is where environmental problems are managed and workers are not being exploited – but this is not always the case in the poorer countries of the world.

The problems faced by poor countries

Exporters of seafood from developing countries face two issues. First is the situation whereby the production methods used in their country are deemed to be environmentally destructive by potential consumers in the rich countries. Note that it is unlikely that the rich-country consumers will have health or environmental concerns on their minds when they buy seafood unless there has been a publicity campaign directed at this very matter. An unaware consumer concentrates on price. That stated, in this era of general environmental awareness, it wouldn't take much effort to influence consumers to be worried about imported seafood. As Jeffrey Frankel states, when environmental degradation is at stake, 'the case for countries sticking their noses into each other's business is stronger than for economic interdependencies'. As he says, 'we all share a planet'.[12]

The consumer complains

It is not only foreign fish-farmers who are the targets of rich-country seafood consumers. In the wild-capture fisheries the main problem is the incidental catch of 'icon' species – endangered or, simply, 'loved' species are unintentionally caught along with the target species. This presents a real dilemma: one country's loved animal can be another country's food.

The peoples of countries such as Norway and Iceland, and the Inuit people of North America, have relied on whale meat for millennia, just as the Australian Aborigines relied (and traditional ones still do) on kangaroo. Today, kangaroo is eaten by European Australians and their pets (dogs and cats), yet it is an animal that graces the Australian currency and the national emblem. Yet some Australians object to others eating whales. There is little consistency in icon animal campaigns. I note Peter Singer's strong argument against killing whales. However, I believe that Singer struggles, as most of us do, about where to draw the line. As the English Prince Charles is reported to have said, even plants scream when being harvested for food. On this issue, scientists cannot take sides – but they can ask for consistency (logic) in the argument.

On to more everyday concerns. The bitter argument between the USA and Mexico (going back to 1990) over dolphins being caught by Mexican tuna boats is a case of conflict over harvesting methods and differing attitudes as to what is important. In 1990, the USA placed an embargo on the import of tuna from countries not meeting US standards on incidental catches. The USA went to a General Agreement on Tariffs and Trade (GATT) – now the World Trade Organization (WTO) – dispute settlement panel but the conflict was ultimately dealt with by a mutual agreement between the antagonists. While the details of this case would be of interest to those wishing to explore the workings (and shortcomings) of the GATT dispute resolution apparatus, this economic intrigue won't hold us up here. However, the point needs to be made that many years later the WTO process, and the institutional arrangements, are still lacking sufficient rigour to allow for certainty when externalities (such as incidental catches) form part of the debate over free trade.

Not knowing when you have won

To Frankel again, who argues that, 'the WTO has actually moved importantly in the environmentalists' direction in recent years … [but] the environmentalist community has almost entirely failed to notice this major favourable development … by ignoring their victory … environmentalists risk losing the opportunity to consolidate it. Some players, particularly poor countries, would love to deny the precedent (favouring the environmentalists) …'[13] He makes a very good point – winning the war, not the battle, should be the game plan.

Let us not assume we live in a rational world, believing that emotion should not override science in the whaling dispute. We are a long way from that. Admittedly the Japanese whalers do a serious injustice to the definition of science by asserting their catch of whales is purely for scientific purposes. One can be one's own worst enemy!

Solving the incidental catch problem

In recent years there have been important and significant improvements in fishing technology (for example, devices that extrude captured animals such as turtles) and, as a consequence, incidental catches have been greatly reduced or eliminated. In response to environmental pressures, the use of selective gear (nets, pots or lines that specifically target a species and deliberately reduce incidental catches) has become commonplace.

However, while focusing on addressing the incidental catch problem, we must not relax our efforts to maintain sustainable catches of target species. Fisheries' scientists have the difficult task of establishing total allowable catches. Scientists can't walk around and count the number of fish in the sea in the way that they could with trees in the forest or corn in the field. Considerable scientific effort is involved in estimating sustainable harvests. Once we have done the science, fisheries' managers need to ensure that the scientifically established sustainable catches are not exceeded. This requires patrolling the oceans and monitoring catches.

The responsibility falls on fishers

The major responsibility for the future of fishing will fall on fishers, if they value their future livelihoods. If they don't, the public perception towards so-called uncaring exploiters of the seas will lead governments to wrong decisions and the closure of many, maybe all, fisheries. Fishers will need to change practices and use gear (nets, lines and such) that do no harm to non-target species or the environment in general, and they will need to adopt codes of practice that satisfy consumers. They are adopting these principles gradually.

There is a key role for the consumer in this new world of sustainable fishing. The labelling of seafood can assist consumers to become better informed: 'dolphin safe' is now a common message on cans of tuna – the tuna apparently being caught without incidental catches of dolphins. We should extend this concept to all major seafood species. Of course, to take hold, claims of environmental responsibility need to be certified by third parties. This is the type of world we live in – the advertisers' exaggeration, the peddlers' push and pull, and the shysters' spin have made us ever so wary. This is an enormous shame and causes loss of social capital. The proverbial handshake is no longer good enough in many parts of the world – sadly!

The trade problem

There is a second part to the environmental quality and international trade problem, and this one is difficult to resolve. Fishers in high-cost developed countries believe they are facing unfair competition from low-cost developing countries. In this case, aquacultured species, particularly shrimps, are likely to be the cause of the conflict. As already noted, shrimp from Asia is available in shops in the West at one-third the price of product from high-cost developed countries. That is an enormous advantage if price is the only factor a consumer takes into account.

Notwithstanding the slow freeing up of world trade through the efforts of the WTO, and the attempts to separate genuine environmental concern from 'backdoor' trade protection by high-cost competitor fishers in the industrialised countries, the issue of real or perceived unfair trade in seafood is very much alive. The price differential, of the order just mentioned, is the source of the problem. Is the difference in price purely the result of very cheap labour in poor countries or lax pollution and health laws permitting low production costs? Maybe it is a mixture of both. Rather than deal with assertions, the facts have to be ascertained before a solution can be sought. The answer is not obvious because the facts are hard to discover. Claim and counter-claim, both with supporting data, confront the neutral umpire.

It is not just seafood that is caught up in the fair trade agenda. Virtually any product from a developing country can be subject to claims that it is underpriced due to sweatshop labour and inadequate environmental controls. As a step forward, there is scope to allow consumers to solve the problem by making them better informed – something I have already advocated. In developed countries, labelling of products is either done voluntarily by producers or is required by law. The most basic labels identify where the product (seafood in our case) is sourced: is it of domestic origin or imported? Consumers can choose between a lower-priced product which might – or might not – have been grown in ponds cut into mangrove forests or have caused pollution offshore and the much higher priced product which has had to run the gauntlet of strict environmental laws in a rich country. Country-of-origin labelling is a crude measure and is far from the answer to the problem we are discussing. It can work against a top-quality Third World supplier – and such supplies do exist. Something better than a simple label has to be provided. Some sort of accreditation system is required as a means of verifying the authenticity of the label. A small number of such accreditation programs exist and there is one specially developed for fisheries by the Marine Stewardship Council. There is, however, little recognition of the need for this in poor countries and the accreditation program has a very long way to go in gaining acceptance if it is to be a major part of the solution.

Assuming we made progress in identifying clean, green seafood products there are other matters that could undo our best attempts at initiating both fair and free trade in seafood (or anything else). In putting the solution in the hands of consumers are we in danger of assuming too much about the rationality of human behaviour? To what extent can we rely on 'green' consumers to help solve environmental problems? In the first instance, we don't know how many consumers are 'green'. Are we assuming that such consumers' concern for the state of the environment in South-East Asia is strong enough to outweigh a dramatic price differential in the price of prawns in their supermarkets – prawns that are going to be thrown on a barbeque or drowned in a curry sauce!

ENSURING SUSTAINABILITY

There are other questions about seafood sustainability: how resilient is the ecosystem in which the fish live? How do we ensure that the interrelationships between fish and other marine animals reach a new sustainable equilibrium once fishing becomes a stable part of ecosystem dynamics? Relationships change, but new ones form and sustainability is possible.

Australia leads the world in tackling these difficult biological and ecological questions. It has a law that requires its fishers to meet extremely strict ecological criteria before they are given permission to export their product. If the same type of laws were put in place in other countries, including developing countries – and if these laws were enforced – this would go a long way to guaranteeing the sustainability of the world's fisheries. The Australian government has taken a principled, but potentially costly, gamble by imposing such strict environmental laws on its fishers. The Australian fishing industry is high cost and environmentally friendly. However, Australian fishers will only be able to compete globally if their enviromental credentials compensate in the marketplace for the high cost of their product.

We should not get too optimistic about reforming fishing practices on a global scale. It is an enormous challenge to harmonise laws of any kind across nations. The EU has shown it can be done at a regional scale (where all member countries have had to reach minimum standards). In the EU case the political, economic and social environments are not significantly divergent across nations in the way they are in all other regions of the world.

Take a region as diverse in stages of development, in religions and cultures, as Asia, which comprises Japan and other First World countries such as Singapore, Korea, Taiwan and Hong Kong; middle-range developing countries such as Thailand, Malaysia and China; and extremely poor countries such as Pakistan. The task of harmonising laws (of any kind) in this region is in order of magnitude harder than it is in the EU. It is not at all clear that other Asian or South Pacific countries (except New Zealand) will follow Australia in enacting laws to ensure the sustainability of fisheries. Even the environmentally conscious European countries lag behind Australia on fisheries legislation.

Likewise, the cost differences in labour between the poor and rich countries will remain for a long time. First World fishers are not going to work for Third World wages, and Third World fishers have no collective power to force First World wages on their employers or to do favourable deals with middlemen, who through their financial involvement in bankrolling poor fishers, come to control the economic destiny of the fishers. That is where we are at today – a situation in which we can think of rational reforms to the global fisheries trade yet cannot see how to implement them.

Free and fair trade

What we have been discussing is the real or perceived fact that free trade is not necessarily the same as 'fair' trade. As a consequence, a fair trade movement has developed. Fair trade has become a recognised consumer label in parts of Europe, and to a lesser extent in other parts of the industrialised world. The movement attempts to achieve its aims by various means: one is to eliminate middlemen and the added cost they impose on consumers and the power they exert over poor producers. This is easier said than done. Certain activities by middlemen are difficult to eliminate. Someone else has to become responsible for the coordination of production, distribution and marketing if the middlemen are removed. The only alternative is a producers' cooperative. These have worked in the industrialised countries.

What is objectionable about the practice of many middlemen in poor countries is their lending of money at exorbitant interest rates to fishers and farmers. Poor producers find there is no escape from debt if interest rates are 100 per cent and upwards. These middlemen are primarily financiers, not conduits between fishers and consumers, as is the conventional role of the middleman.

The goal of fair trade is to be achieved by creating direct links between small-scale producers in the developing countries and wholesalers in the developed countries. The local middleman, and his very high financial impost, would, therefore, be bypassed. However, there is little evidence of this occurring and one wonders what has to be done to stimulate it. The idea might prove to be utopian and will have to be driven by concerned First World consumers and First World middlemen who can profit from being fair trade entrepreneurs.

Another fair trade concept is that the retail price of fair trade goods is to include an explicit 'social premium' to provide funds for socially desirable projects in the producing countries. If this idea is to take hold consumers will need to be willing to pay the higher price. There is little obvious action on this front as yet, as the idea has not been tested seriously in the marketplace.

Consumers will have to drive fair trade policy. It is as simple as that. Those who identify with the goals of fair trade will have to be willing to pay the premium. There is presently a small, slowly growing market for clean and green products (for example, free-range eggs and organically grown vegetables and fruits); however, this is not the same thing as a social premium. It is early days in the development of the fair trade movement and, hence, it is difficult to ascertain how strong it will become, and how soon. Because seafood products have been in the forefront of trade–environment disputes, they could be a useful yardstick for the success or otherwise of the fair trade movement.

The future of fisheries, whether wild-capture or aquaculture, is, like all economic pursuits, in the hands of both the producers (the fishers) and the consumers. The sustainability concepts we have been discussing are going to be relatively easy to implement in the industrialised countries. The one remaining problem will be what is referred to as 'technology creep', meaning our ever-increasing ability to catch previously hard-to-catch fish due to increased engine power, better fishing gear and more sophisticated electronics. The financial incentive fishers have to catch as much as they can before someone else does will continue to exist unless fisheries' scientists can estimate sustainable yields and convince governments to cap the catch. The science is difficult but this approach is the only realistic way forward.

For most of the world other more basic fisheries management problems will continue for some time. Most developing countries have as yet not undertaken the scientific work to determine what level of catch is sustainable, let alone anything else that requires urgent attention; hence they are in a weak position to manage their fisheries. Sound policies and good laws are necessary, but not sufficient, to achieve the sustainable development goals we seek. Developing countries usually have in place appropriate laws but policing and enforcement are too often missing. This needs to be addressed urgently – it is a problem that aid donors should target. With the appropriate scientific assistance (including determining sustainable food regimes for farmed seafood), many of the poor countries have a real opportunity to play a significant role in feeding the world with seafood, and to profit from doing so.

1   John Micklethwait and Adrian Wooldridge, *The Company: A Short History of a Revolutionary Idea*, Phoenix: London, 2005, p. 45.

2   B Kearney, B Foran, F Poldy and D Lowe, *Modelling Australia's Fisheries to 2050*, Fisheries Research and Development Corporation: Canberra, Australia, 2003.

3   Mark Kurlansky, *Cod: A Biography of the Fish that Changed the World*, Vintage: London, 1999, p. 160.

4   Kurlansky 1999, p. 162.

5   Kurlansky 1999, p. 165.

6   Of course, demand and supply will tend to equilibrium but it is likely to be at prices beyond what the vast majority of poor people who are dependent on seafood for protein can afford to pay.

7   Always look for a dispute over natural resources if you are seeking reasons for a war. Soldiers might believe, because that is what they are told, that they are fighting for their country or for some ideology, but those in command – particularly if they start the war – have other interests: economic ones.

8   See Charles Clover, *End of the Line: How Overfishing is Changing the World and What We Eat*, Ebury Press: London, 2004, for somewhat exaggerated stories of fisheries failing.

9   It is becoming common to aim for maximum economic yield (MEY) rather than maximum sustainable yield (MSY) on the basis that the former is a lower level of catch and effort (assuming a zero discount rate). A real-world problem – even in countries with logbooks and electronic surveillance techniques – is how to calculate sustainable yields.

10 The Marine Stewardship Council is a global organisation that conducts a fishery certification and seafood eco-labelling program to reward sustainable fishing and promote the best environmental choice in seafood.

11 It is not just nutrition (the omega-3 factor) that is driving the increase in seafood consumption. The 'slow food' movement, whereby much attention is given to the slow enjoyment of food and the sharing of meals, is also having an impact on seafood consumption as its exponents are choosing to eat 'slow fish'.

12 Jeffrey Frankel, The environment and economic globalization, in Michael M Weinstein (ed.), *Globalization: What's New?*, Columbia University Press: New York, 2005, p. 129.

13 Frankel, in Weinstein 2005, p. 131.

3

THE BEST DEAL IN NET ENERGY: SUNNYSIDE UP

Civilization as we know it will come to an end sometime this century unless we can find a way to live without fossil fuels.

David Goodstein[1]

The end of oil is a reality. Geologists can debate the timing, but the rest of society has to make a decision today that will see us through to the new era of renewable clean fuels. The end of oil is both a problem and a godsend. It is a problem because industrialised society has come to depend on oil and its derivatives. It is a godsend because we cannot afford to continue to fill our skies with greenhouse gases. We have already overdone it and face extremely difficult decisions about how to reduce the amount of the gases being emitted, and adjust our lifestyles in those places (for example, low-lying Pacific Island countries) where sea-level increases are a threat to their very existence. As a first step in understanding our problem, let us start by summarising the end-of-oil story. There is extensive literature on this subject and it is necessary to only present the briefest sketch to underpin our story.

THE END OF OIL

In the early days of the oil economy, when large shallow oilfields were being exploited, the amount of energy required to obtain 100 units of energy was one unit. This was a 100 to 1 ratio – a real windfall, the payout of a lifetime. As smaller, deeper and more remote fields came to be exploited, the net energy output fell. By the 1970s it was a 23 to 1 ratio.[2] That was nearly 40 years ago. Today there are oil wells where we use more than one litre of oil to obtain one litre.

Figure 3 shows the history – and the future – of oil production. The graph represents what experts call the Hubbert peak. The actual peak, the top of the pyramid, is likely to be reached any time soon and, sometime in the twenty-first century, we are likely to 'run out' of oil. There will still be oil in the ground but it will take more than one litre of oil to extract one litre and that is when the oil economy is finished. A former Saudi oil minister, Sheik Yamani, describes this.

> *Thirty years from now there will be a huge amount of oil – and no buyers. Oil will be left in the ground. The Stone Age came to an end, and not because we had a lack of stones, and the oil age will come to and not because we have a lack of oil.*[3]

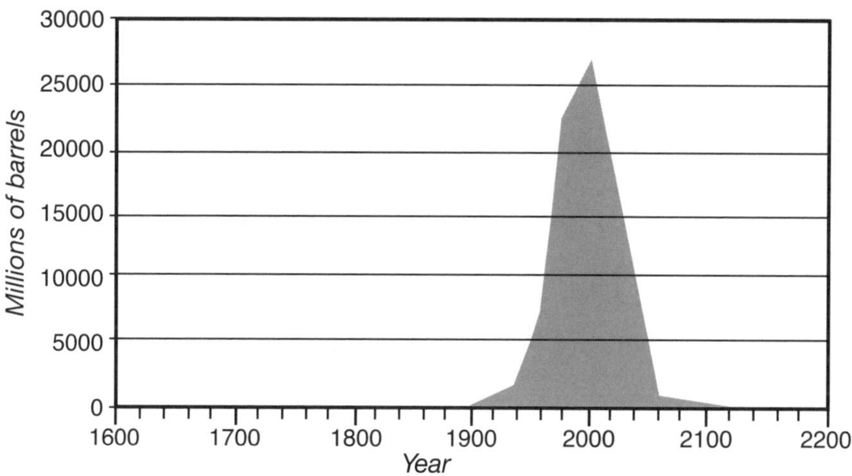

Figure 3: World oil production from 1600 to 2200, history and projections in millions of barrels per year

A world without oil

Consider for a moment a world without oil. Consider how we, in the industrialised countries, have benefited from oil. For example, it would take 2000 people to do the same amount of work that one American car does as it cruises the highway.[4] One gallon (4.5 litres) of oil contains the energy of 50 well-fed slaves.[5] Sonia Shah describes our dependence on oil:

> *Within a century of drilling the first oil well in 1859, crude oil and the products and machines it makes possible had seeped into virtually every crevice of Western society. Today, oil and its progeny make cars run, planes fly, houses warm and lit, hospitals sterile, and supermarkets stocked with fruit and vegetables. Strapped into steely, oil-fed motors, soft, breakable human bodies gloriously extend their reach and power. Newborn babies slide from their mothers into gloved hands, are swaddled in petropolyester blankets, and hurried off to be warmed by oil-burning heaters. The few metal and wood products that remain today are extracted by oil-powered machines travelling on oil-covered roads. They arrive packaged in oil-made plastics.[6]*

Was there a world before oil?

Before oil there was coal, before coal, water-power and before that, for at least 750 000 years, the wood fire kept us warm and cooked our food. Of course, we still use coal, and in extremely large quantities. Coal is fundamental to the rapid development of the Chinese economy as it roars along at over 10 per cent economic growth per year. The conventional wisdom is that it was the decision of the British Navy to convert its fleet from steam to oil power, before World War I, that resulted in the oil revolution. The necessary technological breakthrough that allowed the oil economy to develop was the invention of the internal combustion engine.

We knew of oil from ancient times, from the places where it seeped out of the ground. The Greeks used it to flare arrows. In some societies it was used as a medicine. In recent times its major use, before the internal combustion engine, was for lighting. Before we obtained oil from the ground, whale oil was the conventional means of lighting homes. One wonders what level of protest about whaling there would be today if the lighting for our homes and streets was provided by whale oil. Over-harvesting of the whale, driven by increased demand due to human population growth, gradually brought about the demise of whale oil as the major source of lighting. An alternative was needed.

There was kerosene

There was one oil-based product that predated the use of gasoline (petrol), an oil derivative. It was kerosene. It was a significant improvement on whale oil and other sources of lighting, such as the flammable turpentine used in household lamps.[7] Consider the fact that kerosene comprised only a fraction of the crude that was available, and that the stuff we use today in our vehicles was treated as waste – just as until recently natural gas from oilfields was flared-off as waste. The notion of what is a waste, even the use of the word 'waste', is a problem in our search for sustainability.

In 1879 the light bulb was invented and the kerosene lamp was to disappear in the rich cities of the world – but not everywhere. It was the only household lighting when I was a young child in the Australian countryside. Today there are still many parts of the world where kerosene lights are the norm. While we take for granted the glare and glitz of our electricity-saturated cities, approximately 24 per cent of people in the world do not have access to electricity.

Come the car and fertilisers

In 1885, Karl Benz attached a motor to a tricycle. The motor car, as we know it, was not far off. In 1893, two American bicycle mechanics designed a motorised vehicle. It wasn't going to be long before Henry Ford commenced the mass production of motor vehicles. Note the first car engine could have been fuelled by a bio-fuel: peanuts were proposed! How different the world could have been – imagine enormous fields of peanuts across whatever country they could be grown.

It wasn't just the motor car that came to depend on oil. In 1909, it was discovered how to use petroleum power to make ammonia and we were on the road to artificial fertiliser. Nitrogen fertilisers changed the world, as the list of facts in Box 1 show. This is how Shah describes the discovery:

> *Employing methane to create extreme heat (up to 600 degrees Celsius) and intense pressure … Haber transformed nitrogen from air into NH_3, ammonia, which could be used to make nitrogen fertilisers. Petroleum had allowed Haber to capture the Holy Grail of inorganic chemistry.*[8]

Box 1:
Nitrogen fertilisers change the world

- Agricultural yields doubled between 1947 and 1979. This was the period of the so called 'green revolution' in Third World agriculture.

- Population skyrocketed during the same period.

- Before nitrogen fertilisers there were 2 billion less people in the world. Without nitrogen fertilisers, the world's population would have stabilised at 3.6 billion; that is, 3 billion less than we have today. (We will need to feed an extra 3 billion before the population plateaus at 9-plus billion).

- Large mechanised farms rely on oil for fertilisers, pesticides, harvesters, transport (road, sea and aircraft, the latter for the high-value-added products flown from distant markets).

The world is plastic, isn't it?

Then another use was found for oil: the manufacture of plastics. Plastics have become the most utilised materials in the world: they are found in glossy paper and printing ink, vinyl floors, Teflon pots and pans, all sorts of clothing, artificial limbs, children's toys, a vast array of wrapping, and motor vehicle and airplane parts. Today only the ping-pong ball is made from a natural plastic, and the cornstarch shopping bag substitutes for the ubiquitous plastic bag. Litter clean-up campaigns keep most of the plastic bags off the streets of First World countries, while they clothe the fences and, if they escape these, the fields of developing countries.

Shale oil

So far we have confined ourselves to conventional oil, or light crude. There are a number of other sources of oil: oil sands, tar sands and shale oil. Exploiting these will be more expensive, more environmentally damaging and slow to develop, if they ever are. As we move from light crude to shale oil, the net energy (the oil input to output ratio) will increasingly get worse, until the input of oil exceeds the output of oil. In fact, some argue that exploiting shale oil will never produce net oil benefits.[9] The same, by the way, is said (by some) to apply to corn-based ethanol in the USA.[10] This is disputed, as much depends on the scale and type of corn farming; that is, whether it is energy or labour intensive, irrigated or rain-fed. The energy-out for energy-in ratio is improved the less energy intensive the growing and harvesting of the grain. Using waste from corn production also makes a big difference.

Natural gas

After oil is finished there is a bridging source of energy – natural gas. By bridging, we mean something to see us through to the full-scale uptake of renewable fuels. Natural gas (mostly methane) provides the most likely alternative in the short term. It is easy to extract. The transformation to a natural-gas-fuelled economy would be made much more easily than to other alternative fuels, in particular because the normal car engine can be run on compressed natural gas. However, the fuel distribution system would need to be radically changed if universal use was to be made of natural gas. Large-scale plants to convert methane to gasoline would take some time to construct; hence the timing of conversion from oil will need to be planned carefully. And, of course, there is a Hubbert peak for natural gas. It will probably come only 20 years after the peak for oil – which is not far away. However, it might just allow enough time for the required radical transition to renewable fuels to take place.

Back to coal

Could we go back to a totally coal-powered world (recognising that some still use coal)? Coal is elemental carbon. There is an enormous amount of potential chemical energy stored in the earth. As with all fossil fuels, to extract stored energy from coal, each atom of carbon has to combine with oxygen to become a molecule of carbon dioxide – the major greenhouse gas. Coal also contains a range of impurities – sulfur, arsenic and mercury – all quite expensive to remove. Essentially, 'coal is a very dirty fuel'.[11] That stated, there are estimates that we have enough coal in the world to last for hundreds of years. But, if we factor in the rapid economic growth of China, and later India, and their dependence on coal, we start to quickly contract the number of years of a coal-fuelled world. There also is, as we have come to recognise, a Hubbert peak for coal.

The big bang!

Back to nuclear power – only a few decades ago it was full of promise as the fuel of the future. It fell out of favour in many countries partly due to its cost, partly due to unfortunate accidents (Three Mile Island and Chernobyl) and partly due to issues with security of supply and the danger of nuclear proliferation in the arms race. Today nuclear power is back on the agenda as another option in a world of diminishing fossil fuel resources.

There are two types of nuclear power. Nuclear fission is long established and there is considerable fuel available – the radioactive isotope uranium-235. Nuclear fusion is not available yet, and may never be. In theory, it is the process of obtaining energy by fusing light nuclei into heavier ones. The fuel is deuterium, a form of hydrogen found naturally in seawater, and lithium, a light element found in many common minerals. I suspect the anti-nuclear lobby is not powerful enough to stop a revival of the nuclear-power industry – particularly if there are no feasible, cost-effective alternatives. The search for such alternatives is in full swing in some countries, while others oscillate and procrastinate.

FUTURE SCENARIOS

Having put to bed, so to speak, the problem fuels – and before we consider the environmentally friendly alternatives – let us think of some future scenarios. They will set the scene for what we need to achieve and when. There is, as always, a worst case. This is where we fail to bridge the gap between the 'end' of oil – more realistically, the Hubbert peak – and an alternative energy source that is fully on stream. The consequence will be runaway inflation, leading to a worldwide depression. In this case, the only large-scale available fuel would be coal and it would need to be used in vast quantities for household needs (cooking and heating) before any would be available for industry. The considered scientific view is that the greenhouse consequence of complete reliance on coal would totally tip the world climate, and life as we know it – or, more dramatically, any life – would disappear. This is, admittedly, a dramatic scenario. It ignores the possibility of so-called 'clean coal'; that is, coal from which carbon dioxide is extracted before being released into the atmosphere. It also overlooks other possibilities in a world of nine billion plus people competing fiercely for the available coal – war or the collapse of the world economy, or both. Before we despair at this prospect we should ask another question. Could we return to the life we lived before widespread use of coal, as it was 200 years ago? There are utopian dreamers who tend to think so – this is wishful thinking of the highest order. The answer is found in one simple number. We would need to rid ourselves of 95 per cent of the human population, as David Goodstein points out in his 2004 book, *Out of Gas: The End of the Age of Oil*. Forget this as an option.

The best possible outcome

From one extreme to the other: the best-case scenario is one in which we take immediate notice of the impending peak in oil production, and act now. Natural gas (methane in particular) would become the bridging fuel and be used extensively throughout the world economy. But this is not the only solution. For some analysts there is another option: nuclear power would become popular again and new plants would be commissioned. As the Hubbert peak for natural gas approaches, nuclear power would be in a position to take up the slack. The 20 or so years of full-scale natural-gas production would provide just enough lead time to commission and begin operating nuclear plants. However, the fear of nuclear disasters would remain, and the safe renewables would gradually gain a foothold as technological progress and, extremely importantly, the necessary distribution infrastructure for their use was developed. This means that as the Hubbert peak for uranium was approached, the safe renewables would take over. Maybe the nuclear part of this scenario won't occur. The unresolved matter of dealing with nuclear waste suggests the nuclear option will continue to be a country-by-country decision. Let us make the point once more: the twenty-first century is 'the century in which we must learn to live without fossil fuels'.[12]

Electric batteries and hydrogen

Given our reliance on motorised transport in our much-cherished modern lifestyle, developing a new fuel and an efficient distribution system is one of the most difficult problems we face. A number of possibilities sit between the two extreme scenarios – or are components of a modified best-case scenario. One is the development and use of an advanced electric battery – think of the tiny batteries used in mobile phones and laptop computers. Then there is hydrogen – ordinary hydrogen – burned by simple combustion, or used in hydrogen fuel cells that produce electricity directly. Both these methods of using hydrogen release nothing more than water vapour into the atmosphere. While water vapour is a greenhouse gas it is quickly cycled out of the atmosphere as rain, hail or snow.

We should note that – thermodynamically speaking   hydrogen and batteries are not sources of energy, but rather are nothing more than a means of storing and transporting energy. Where does the energy come from to make hydrogen? Coal is one answer. You can combine coal and steam, but, of course, the process makes carbon dioxide. This could be sequestered – somewhere; however, coal is not a long-term solution as we have noted. Both sunlight and nuclear energy can be used to make hydrogen from water. In a nutshell, hydrogen or advanced batteries can meet our transportation needs – if we make rapid technological progress and, just as necessary, put the infrastructure in place. These are big asks given the fact that the urgency of replacing traditional energy sources is not recognised by those who could do most to address the issue: the financial sector, the major players in oil and national governments.

SHOCKED INTO ACTION

The genuinely 'new' renewable energy sources are in the state that petroleum was 100 years ago (not that long ago), when oil and its derivatives had uncertain futures. Today the new fuels account for only 2 per cent of energy use worldwide. The Organization of the Petroleum Exporting Countries (OPEC) oil price increases in 1973–74 and then again in 1979–80 caused the first two waves of interest in the new fuels. Little happened – except in university laboratories. A notable exception was in Brazil, which commenced its very successful trajectory into an ethanol economy.

It has taken two recent phenomena to turn those initial small waves into a potential tidal wave. First is the very high price of oil and the predictions that, this time, it won't fall as far as it did after the first two oil shocks. The Hubbert peak is real! Second is the acceptance that global warming is occurring due to the build-up of fossil-fuel-generated greenhouse gases.

Let the sun shine!

Let us continue with the discussion of possible fuel choices. If not fossil fuels or nuclear power, there is only sunlight. Humans have used sunlight for as long as we know. Sunlight is needed to grow trees; trees provide us with the wood we burn for warmth and to cook food. We use wood in massive quantities today for the very same purposes that we always have. The ubiquitous charcoal burner cooks food in the streets of undeveloped countries throughout Asia. Wood fires burn in every African village. This will continue for some time. Population increase puts serious pressure on limited wood supplies (an increasingly common Third World problem) and the only feasible reaction is to plant trees.[13]

Today, sunlight 'technology' is no longer charcoal but the solar cell. Solar cells convert sunlight directly to electricity, but solar radiation is diffuse (not concentrated) and is not easy to store. Storing it requires conversion to some other form of energy – such as heat, electricity or chemical energy. What are the limitations? How much of the earth's surface could we feasibly cover with solar collections? We could cover all existing rooftops – this would be feasible if the need arose. It has been estimated that it would be possible to meet the total US energy requirements if every centimetre of roof area in that country was covered with solar collectors. This project would have to include all commercial buildings in addition to homes. Possible, yes; practical, no – unless it is the last straw.

Solar photovoltaic cells have declined in cost in recent years and there was strong growth in their use worldwide in the 1990s and into the twenty-first century. Three-quarters of the world's countries now have photovoltaic cells in some form of use. BP Solar (an offshoot of the fossil fuel company) is the world's largest manufacturer of cells and had a one-fifth share of the world market at the beginning of the twenty-first century. The company manufactures cells in the USA, Spain, India and, until recently, Australia. Use was initially confined to remote (off-grid) areas. Now that subsidy programs exist (for example, in Germany, Japan, the USA and Australia), grid connection is becoming popular for large photovoltaic producers.

Direct solar energy is certainly likely to play a much greater role in the future, but to what extent is yet to be determined. Let us consider some of the indirect solar sources of power, as they offer another range of possibilities.

Hydropower forever

As we know, the sun causes water to evaporate and to later fall back to earth as rain or snow. And then the rivers flow. Any country with an abundant quantity of flowing water has a hydroelectricity capacity. This is not a new technology – it only seems new in those countries where it has been neglected for ages or simply deemed too politically difficult to implement. The two related matters of peak oil and climate change have rekindled interest in hydropower, particularly in small-scale units. To be fair, we should note that even the very small-scale or micro-hydropower generators we hear much about today have been used for a long time. My grandparents' home in northern Norway obtained its lighting and heat from a very small hydro-generator installed in the little but fast-running creek that was fed by the snow on the nearby mountains.

Hydropower is very simple at the large scale: dams are built and water stored, then to be released. The enormous pressure of water in a dam – or simply water falling in a fast-running stream – provides the force to drive a water turbine. Electricity is generated. This process can be organised to provide electricity day in, day out, year in, year out. Norway runs its whole industrial economy on this source of energy. Iceland complements its geothermal power with hydropower. Canada has abundant hydropower. It is found in all continents – even Australia, which is the driest continent after Antarctica. Still, in the major world economy (ie in the USA) only about 10 per cent of electric power is generated by hydro sources.

Limits to hydropower

There are two problems if we wish to extend the use of hydropower. First, we have dammed most of the available sites throughout the world. Finding large new sites is very difficult. However, there are tens of thousands of small-scale sites yet to be utilised. The second problem is that dams have their own social and environmental problems. In heavily populated countries, such as China or India, millions of local people have to be forced to move to make way for a mega-dam. The loss of biodiversity can be considerable, depending on the extent of prime natural land that is inundated. The Three Gorges Project in China and the Narmada Dam Project in India are very controversial due to their impacts on local farmers and the ecosystem.

There is also the fact that dams eventually lose their holding capacity due to siltation. On the positive side, there is considerable scope for micro-hydropower in remote areas where fast-running streams exist. But remember that for large schemes there is a scarcity of appropriate sites, and the annual growth rate in hydropower during the 1990s of just under 2 per cent indicates this problem.

On the crest of a wave

There are various sources of power to be derived from the oceans. Ocean currents, of which a good example is the Gulf Stream, carry enormous amounts of kinetic energy. Ocean turbines could be placed in the strongest currents and the generated energy transported to shore. The downside of projects such as this is the enormous cost involved in constructing and anchoring huge structures in strong currents in deep water. Oil and the other alternatives will need to become much dearer before such major offshore projects will be feasible.

There is also the thermal energy in oceans. The surface water of tropical oceans is much warmer than water at great depths. A heat engine could be constructed to operate between these two thermal regions of the oceans, but again the cost is a serious impediment at present. But always be mindful of the fact that the cost of an energy source is relative to that of its competitors. At present, these ideas about utilising these forms of ocean power will remain just that – ideas.

Tidal power can be obtained by placing turbines in or across the mouths of estuaries that have significant tidal variations. Energy from the flow of water in and out of the estuary is extracted and turned into electricity. This is a far more realistic concept than the previous two. There is a tidal power plant in France in the Rance River, near St Malo, Brittany, which has been in use since 1966 and supplies in the order of 90 per cent of Brittany's electricity.

The waves breaking upon our shores carry large amounts of energy. Devices presently exist that can extract energy directly from waves at the surface or from pressure fluctuations below the surface. Nevertheless, the oceans as a source of energy have not been elevated to the status of a major competitor with other energy sources, mainly due to the cost involved in the serious engineering that is required.

Wind in the willows

Wind power is another indirect form of solar power. It is becoming increasingly important in northern Europe, particularly in offshore, coastal or hilly locations that are subject to constant winds. Wind power experienced a fantastic near 25 per cent annual increase during the 1990s – but this is from a low base. Due to the high taxation of fossil fuels in Europe, and the electricity in-feed laws with a guaranteed price paid to wind-farm businesses, wind power is an economical competitor with traditional energy in many places in Europe. On the other hand, only one-quarter of 1 per cent of the USA's electric power comes from wind.

Wind power (like hydropower) is limited by geography. This lessens its appeal. There is, furthermore, in some countries, opposition by local people to wind farms. This is what we call the not-in-my-backyard (NIMBY) syndrome. They don't like the sight of the spinning blades and towers, or the noise, yet there is no objection to using the electricity generated. When the world runs out of oil and there are few substitutes, this opposition is likely to abate. Where there is no or only little opposition, wind power continues to grow. Within the next decade, wind 'parks' – mainly offshore – will provide the energy equivalent in capacity to several large coal-fired or nuclear power stations.

The major countries utilising wind will be Denmark, Spain, Sweden, Germany, Holland, Belgium, Britain and Ireland. Yet the east coast of the USA is a suitable area, as are various parts of southern Australia. There is no practical reason why wind farms on land should not be a component of conventional farming, providing another income source. Plants can grow under windmills and cattle and sheep can graze around them – all very positive.

Geothermal power

Iceland escaped any requirement to reduce greenhouse gases in the Kyoto Protocol in 1997 because it runs its economy on geothermal power and hydropower. Geothermal energy, from the interior heat of the earth, can be accessed anywhere on earth. It is being developed slowly in some countries. It grew at about 4 per cent, on average, over the 1990s, but in most places it requires drilling that is too deep for it to be a feasible energy source. Only in certain geological circumstances is the source close enough to the surface to be viable, given the cost of competing energy sources. That stated, geothermal sources must be considered a serious replacement for fossil fuels in the not-too-distant future. No complex technology is required and the resource is enormous: 'The entire world resource base of geothermal energy has been calculated in government surveys to be larger than the resource bases of coal, oil, gas and uranium combined'.[14] Geothermal energy could be the surprise winner in the future energy stakes. It could leapfrog its competitors and become the number one source if governments and large financial investors become interested.

Biomass energy

In its traditional form, biomass (as wood fires) accounts for less than 15 per cent of world energy. However, this situation won't last. Today biomass is increasingly being used in modern applications as an energy source, drawn from a variety of materials. Many new applications of biomass energy are under consideration. The United Arab Emirates (discussed in Chapter 7) are presently oil-rich, but the Hubbert peak approaches rapidly and the country is seeking to sustain its income by investing in developing energy technologies as well as diversifying its economy. The thinking of decision-makers in the UAE is such:

> *Biomass has tremendous potential to supply future energy requirements, with many different possible sources and conversion methods. Energy crops (such as short-rotation forestry) … starchy plants (such as tubers) and sugary plants (sugar cane) can all be converted into energy. The former are best burned directly to provide heat or to drive turbines to produce electricity, though they can be gasified or pyrolized*

in low-oxygen processes to produce syngas or liquid fuels. These fuels can be used for combustion for heat and power, or in transport applications. Other forms of biomass include … animal (and human) wastes, which can equally be burned directly or processed into gases or liquid fuels. While some care must be taken that energy crops are not grown as a substitute for necessary food crops, it is often possible to use the residue of the latter as an energy source or to use crops from the areas that are unsuitable for growing food.[15]

In some areas, particularly with tropical climates, where the growth rates of plants are fast and the growing seasons long, there is a real attraction to biomass energy. Sugar cane, as an example, can be grown both for food and energy production (in the form of ethanol), as it is in Brazil.

Making the best use of sugar cane

It is possible to illustrate the benefits of converting an economy (in fact, a whole region of the world) from reliance on fossil fuels to a renewable energy economy using the example of the country Fiji and the region the South Pacific. Fiji has an extensive highly subsidised sugar-cane industry. From being the mainstay of the Fijian economy not that many years ago, it is now under serious threat due to international trade liberalisation (the end of subsidies) and a land-tenure system that constrains the industry to small-scale inefficient farms worked on a leasehold basis by Indian Fijians. Without significant institutional changes – and such are unlikely – the continued production of sugar cane for food is not sustainable in Fiji.

However, there is salvation for Fijian sugar-cane farmers. Brazil has shown the way by developing ethanol from sugar cane to power its motor vehicle fleet. It took 30 years of consistent adherence to an ethanol policy, through the ups and downs of crude oil prices and fluctuating sugar prices. The Brazilians must have often been tempted to abandon their commitment to ethanol, but they have proven the technology and stuck with institutional arrangements required to make their energy policy work. Brazil has illustrated that the essential requirement is the political will to stick with ethanol, regardless of any major fluctuations (importantly, decreases) in crude oil prices.

Every country in the South Pacific is dependent on fossil fuels for transport, industry and domestic use. Fiji is somewhat of an exception as it has significant hydroelectricity resources, but as a highly populated country with relatively high car ownership, fuel is a significant (and costly) import. Some years ago, in 2004, Fiji reached the stage of importing over 500 million litres of petroleum fuels (excluding aviation gas).[16] The cost was F$279 million (US$187 million). The use of this fuel generated over one million tons of greenhouse gases in that year. Extrapolation into the near future (based on population growth) shows that both demand for fossil fuels and the emission of greenhouse gases will increase sharply in Fiji. For example, using a realistic estimate of the increase in consumption of 5 per cent, in 20 years time three times more fossil fuel would have to be imported into Fiji than is at present. Not only will increased demand affect the import bill, so will the expected increase in the price of oil. At US$70 a barrel of crude, by 2010 the cost to the Fijian economy for the import of fossil fuels would be three times its present bill. A price of $70 per barrel is not a high price in the present era (given the price exceeded US$100 per barrel in 2008) – and 2010 is very close. Price is one thing – more importantly, there will be an increase in greenhouse gases emitted.

There is a relatively simple solution to the twin problems in Fiji of an escalating cost of petrol and diesel and increased greenhouse gases: use sugar cane to make ethanol. Because sugar-cane production in Fiji is labour (not energy) intensive and rain-fed (not irrigated), the net gain in energy is substantial. It would possible to phase this in by commencing with an ethanol–petrol blend; however, the sooner the better to go to 100 per cent ethanol. The cost to upgrade the sugar mills in Fiji to produce ethanol is estimated to be F$150 million (US$100 million). For that investment, Fiji would rid itself of a large and growing fossil fuel import bill. It would also be able to export pollution-free fuel to the whole South Pacific. This is the regional solution.

An investment in ethanol production presents a wonderful opportunity for aid donors. The cliché of a win–win is a reality in this case. Yet the political uncertainty in Fiji – there have been four coups since independence at the beginning of the 1970s – puts a serious dampener on any proposal for an ethanol-based economy.

The high moral ground

To another small island. Iceland will soon have the world's first hydrogen-powered vehicle fleet, while the rest of its economy will rely on geothermal and hydropower sources. Sweden is in line to have a non-fossil-fuel economy at about the same time – well within the next decade. To see Fiji (and the South Pacific) up there with the high-flying Scandinavians would be a source of pride for the local people, as well as a drawcard for the increasing number of 'green' tourists. The contribution the South Pacific islands would make to reducing the global total of greenhouse gases would be insignificant, yet what better way to influence world public opinion that its low-lying coral atolls should not have to surrender themselves to the destruction caused by sea-level rises. It might be too late for the lowest of these islands, yet the Pacific region will have taken the high moral ground if it makes this fuel conversion – and this can only help it obtain the required foreign aid when it becomes necessary.

1   David Goodstein, *Out of Gas: The End of the Age of Oil*, WW Norton: New York, 2004.

2   Hydropower has an energy return of 11 to 1; coal has a return of 9 to 1 and nuclear power has a return of 4 to 1

3   Sheik Yamani, in *Daily Telegraph*, June 2000.

4   Richard Heinberg, *The Party's Over: Oil, War and the Fate of Industrial Societies*, New Society: Gabriola Island, Canada, 2003, p. 30.

5   Heinberg 2003, p. 30.

6   Sonia Shah, *Crude: The Story of Oil*, Allen & Unwin: Crows Nest, NSW, 2004, pp. 3–4.

7   Shah 2004, pp. x–xi.

8   Shah 2004, p. 5.

9   Shah 2004, p. 18.

10   Goodstein 2004, p. 32.

11   Goodstein 2004, p. 32.

12   Goodstein 2004, p. 33.

13   Of all the things that could have taken me to Nepal more than 20 years ago the most unlikely was to report on an Australian aid-funded social forestry project that aimed to address the diminishing source of fuel wood. The rural Nepalese peasant people have for ages stripped trees for fuel for their sunmestic needs in the cold winters of the Himalayas.

14   Ross McCluney, in Andrew McKillop and Sheila Newman (eds), *The Final Energy Crisis*, Pluto Press: UK, 2008, p. 168.

15   Paul Aarts, *The Oil Weapon: A One-Shot Edition?*, The Emirates Center for Strategic Studies and Research: Abu Dhabi, 1999, p. 72.

16   Industrial diesel comprised two-thirds of the imports, automotive diesel 17 per cent, gasoline 15 per cent and kerosene 1 per cent.

4

SAVING THE ENVIRONMENT CAN BE FUN: ECOTOURISM

Oh my God, what's that snake sucking on my leg!

English tourist on a rainforest excursion. The snake was a leech.

For reasons that I will discuss next, international tourism is singled out by the governments of developing countries as a key industry by which to accelerate economic growth, or if in dire straits, perhaps simply to pull their countries off the bottom of the ladder. Why this attention to and reliance on tourism? First, tourism is a very large and growing industry. Second, it is a labour-intensive industry. Third, most of the workers in tourism do not need to be highly skilled (think of bar workers, door staff and cleaners) and hence the costs of training are manageable by even the poorest countries. And, finally, many of the world's developing countries have magnificent natural environments or fascinating cultural environments or both, and this means they have the necessary tourist 'product' (to use the industry's jargon) to attract large numbers of overseas visitors.

TOURISM: THE ONLY THING TO SELL

It is possible that, in some countries, tourism is the only thing they have to sell to earn foreign currency. They have no oil, no minerals, no agricultural products able to compete with those of the subsidised farmers of the rich countries, no textile workers to sew the shirts and skirts for export, no trained technicians to assemble white goods, or too few English speakers to organise a competitive call centre – only tourism. Yet there is no guarantee that a developing country will make a success of tourism. The reality is that most poor countries do not know how to use tourism to their advantage and, consequently, opportunities to develop this industry and the wealth of the country go begging – while wealthy tourists spend their money in Rome, London, Paris, New York and Sydney – cities which already have more than their fair share of visitors.

Fabulous World Heritage sites

There are over 100 fabulous UNESCO (United Nations Educational, Scientific and Cultural Organization) World Heritage-listed sites in Africa and the Middle East, including in some of the poorest, most debt-ridden countries in the world. By and large, the governments in these areas are not doing enough to use these nature-given and human-made places for what they are – a drawcard (and hence a competitive advantage) to attract international visitors. Of course, it is not simply blindness to the fact that foreigners love nature and are drawn to ancient ruins, old churches and mosques, Buddhist temples and great creations such as the Great Wall of China, to name but a few, that prevents a country making the most of its attributes.

Even when the countries in question attempt to exploit their wonders, other not so trivial matters keep tourists away. Civil wars are endemic to some of these countries, which otherwise would be drawing the crowds. Think of the Middle East with its historical, aesthetic and cultural appeal. Think of the jungles, animals and fascinating cultures of Africa and South America. A civil war or threat of an indiscriminate suicide bombing can suggest too much holiday excitement for most tourists. Then there is the abysmal infrastructure in many of these otherwise enticing tourist locations. Generally tourists have only a few days for their vacation and to waste precious time getting from the nearest airport to their accommodation is a turn-off. Poorly developed and dangerous roads, traffic jams, infrequent transport services or unsafe ferries are major frustrations or cause for worry for the tourist. And there is reluctance for investors to provide money for improvements for a variety of understandable reasons, mainly the lack of effective financial mechanisms by which they will be recompensed. Notwithstanding these impediments, there is a way forward for a nation with a sense of pride in what it has to show visitors. Progress towards a sustainable tourist industry will be made as soon as the guns are put down and bridges erected.

Coming to understand people

There is another reason to focus on tourism in our search for sustainable development. In addition to providing much-needed foreign currency for poor countries, tourism can be an effective means of promoting better understanding between people and, hence, reduce conflict based on ignorance about race, creed or culture. As tourists, we meet people from radically different cultures, religions and races, but we meet and eat at the same table and do so as 'fellow travellers' – in both senses of the term. History has shown us that contact with strangers is one of the great engines of social change. The world's history has many examples of the benefits of cultural exchange. Great civilisations in the Middle East and the Mediterranean – think of those in ancient Persia, Mesopotamia, Greece and Rome – emerged through this process. Always keep in mind that so-called Anglo-Saxon London was fashioned by the ideas, language and infrastructure of invading Romans, Vikings and Normans, as well as the Anglo-Saxons – who themselves were early invaders from Denmark and Germany. The official language of England was French for about 300 years after William the Conqueror became king in 1066! Prior to this, Latin had been the language of high learning. What is the meaning of English, British or Irish?

The tourist experience

People travel for all manner of reasons: for recreation, holidays, to visit friends or relatives, to see new places and to experience different cultures. Tourism planners go to significant lengths to analyse and categorise types of tourists. There is also a very important economic dimension. To be a tourist you must be willing – and able – to spend money. The poor, of course, miss out on the life-enhancing benefits of travel.

Good and bad

Tourism is not just about tourists, it is also about the tour operators, owners of accommodation, travel agents and airline companies (among others) who are working to meet the demands of tourists, and to make a profit in doing so. In the process, employment is created, from high-paying jobs such as resort managers and chefs to the very low-paying jobs in cleaning and garden work. Yet, those who serve tourists cannot but help to learn something – not necessarily something positive – about the tourists' values and attitudes. If workers learn something positive – say they better their English, a step to a higher paid job – this should not be downplayed. If they come to appreciate the fact that the tourists – although enormously and unbelievably wealthy by the standards of the workers – are not 'devils in disguise' but are simply people like themselves, this must be considered a great leap into a better world. On the other hand – and this is, sad to say, just as likely as a positive outcome – if workers discover the tourists to be ignorant overlords, tourism becomes a negative force. Envy supplants any interest in conversing and exchanging ideas with foreigners, who come to be called just that, 'foreigners', in the local language – and usually disparagingly.

THE WORLD'S LARGEST INDUSTRY

Tourism is the world's largest industry, a position it achieved after the end of the Cold War – before then, the armaments industry took pride of place. By the end of the twentieth century, 700 million people had been classified as international tourists (meaning that they had spent at least one night in a foreign country). This is easy for a European who can drive between capital cities, but is a bit more difficult in Asia and the Pacific.

Fifty years earlier, there had been only 25 million international travellers – a figure that increased by 28 times in just two generations. By the year 2020, it is predicted that 1.6 billion people, or one-fifth of the forecasted world population, will be international tourists. They will be looking for interesting – and for many of them, new – places to visit. This presents an enormous economic opportunity for many (admittedly not all) poor countries. It is forecasted that the strongest tourism growth will be in East Asia and the Pacific. Africa, the Middle East and South America will, regardless of their enticements, seriously lag behind in comparison: Africa, because of its ever-erupting civil wars and lack of infrastructure; the Middle East, because it is the Middle East; and South America because it is a long way and seems incapable of, or unwilling to, promote its fascinating cultural and natural environments.

Not all can do it

A country needs the attractions in the first place if it is to cash in on the spending of the rich. Very few countries can create artificial tourism products – the other means of offering a drawcard – that are of significant size and diversity to be attractive to the millions of tourists required if tourism is going to make a worthwhile difference to a poor country's economy. Disneyland is not an option for struggling countries. Countries that have natural and cultural attractions – such as the beaches and islands of South-East Asia and the Pacific – and offer relative safety to the visitor have a great advantage.

THE HISTORY OF TOURISM

Let us explore in more detail what is necessary for a successful tourism industry by stepping back in time. The history of modern tourism covers nearly three centuries. If we wish, we can look back a lot further, as travel writer Tony Perrottet does in his book *Route 66 AD: On the Trail of Ancient Roman Tourists* (see Box 1). In doing so, we come to realise how much things change – and how little things change. Due to this fact, it is not very difficult to predict and anticipate tourist demands and behaviour and, therefore, plan for them.

How little things change

'Modern' tourism started in the early 1700s with the 'Grand Tour' conducted by the sons (with some notable exceptions, women were not in this privileged position) of the northern European aristocracy, as well as the travels of famous scientists such as Carl Linnaeus, who also happened to be an aristocrat. With the formation of a middle class in Europe, as the Industrial Revolution gathered pace, the weather-weary and wealthy sought sun, relaxation and health-giving waters in the Mediterranean in locations such as Nice, the coasts of Italy and Greece, and later ,the Canary Islands.

The weather-weary

Nice – originally an obscure stopover in France on the Grand Tour to and in Italy – promoted its sun as the best in the world, and the rich came to cleanse their bodies, colour their skins and quench their cosmopolitan thirst before returning to the cold and damp of England. As romantic love replaced planned marriages, another reason to visit exciting destinations took hold. The new middle class grew in numbers and wealth and had the means and desire to make honeymoon travel fashionable. This was happening for the well-off by the 1830s. A much sought-after destination to show the new bride or groom (and to talk to envious friends about) was Niagara Falls in North America. Its reputation was such that it was considered the greatest natural wonder in the world. Today waterfalls are somewhat passé but they should not be. That said, we have an extensive list of places that compete for the accolade awarded to Niagara Falls, and those genuine contenders are on UNESCO's World Heritage List (sadly, Niagara Falls does not appear on any 'must visit' list today). In North America in the twenty-first century, the UNESCO-listed favourite sites would include the obvious such as the Yellowstone and Grand Canyon National Parks and the less well-known Head-Smashed-in Buffalo Jump – you just have to love this name!

Box 1:
On the road again: Route 66

The follow-in-the-footsteps genre is an easy way to start writing a travel book. You do your research, as Tony Perrottet did by consulting Lionel Casson's *Travel in the Ancient World*, then you pack your bags. The reader vicariously joins you on your travels and learns a little or much about a culture, people and their modes of travel – past or present.

Perrottet writes that the tourists of AD66, 'stayed at roadside inns, complained about hard mattresses and bad service, ate at dubious restaurants, and got drunk in smoke-filled taverns, and wrote poems about their hangovers … visited lavish temples … handed over hefty donations to shyster Priests for a glimpse of a Gorgon's hair'.[1] Like their twenty-first century descendants, the tourists bought souvenirs – painted glass vials of the Lighthouse of Alexandria and miniature statues of Apollo – or they visited floorshows featuring crocodiles. Not only were these fearsome beasts handfed, but also they had wine squirted into their mouths and their teeth hand-polished. Our modern showmen (including the late Steve Irwin) are not in the same league – otherwise, not much has changed in 2000 years of tourism.

Not all of us, when in Athens, conjure up mental pictures of Socrates, Plato and Aristotle sitting on their speakers' stumps in the agora among the citizenry, who are doing the daily shopping and browsing, and wonder why we have made so little progress in bettering the human condition in 2500 years. The type of thinking the ancient Greeks undertook is no longer fashionable and we are losing the ability to do it. It is nothing more or less than asking, 'How should we live?' It is much easier

> to, while in Rome, have your photo taken sitting on remains of a 100-seat public latrine, with its brilliant system of underground sewers, and ponder how engineering skills have been passed on down the ages. Too many of us don't even do that.
> Perhaps we have become technological giants and intellectual pygmies!

Poets and philosophers stir

It falls on poets and philosophers to stir our aesthetic imagination. Following Jean-Jacques Rousseau's captivating description of mountain streams and abysses came the tourist cult of the mountains, with the inevitable ideal being the Swiss Alps. Today, mountain climbing is high on the scale of 'high adrenalin' (or adventure) tourism. Of less value as a conversation piece (because it is far less dangerous) is trekking. For most of us just having mountains as a backdrop satisfies our desire to be near nature. Henry Thoreau promoted the nature walk, from which modern-day bushwalking and hiking was to develop.

In 1872, the world's first national park, Yellowstone, was declared, and the concept of protecting places of aesthetic appeal and ecological value spread around the world. A country without a formally declared national park is not trying – some completely farm-covered and urbanised small European countries being excused. Australia was not far behind the USA with the declaration of the Royal National Park, on Sydney's southern border, in 1879.

The camera

The invention and popularisation of the camera and landscape photography played a fundamental role in promoting travel, and natural and human cultural artefacts were sought after as backdrops to recreational experiences. This concept grew to the extent that 'entertaining' friends with an evening slideshow of your latest adventure had become very popular by the 1970s. Today, we get from travelling friends regular emails accompanied with digital photographs from mountain villages in Laos or snow-covered peaks in the Alps. This is an easy way to make the point that you are there – while those of us at home slave away at work. And the slideshow has joined other artefacts of the 1970s (such as bell-bottom jeans and the record player) as retro objects, of a monetary value in excess of a recycler's meagre payment for such items.

Blue Hawaii and Baywatch

From the 1960s, the film industry and then television played a significant role in enticing people to dream of sun, surf and sand (you can include sex if you so desire). Popular music by groups such as the Beach Boys and Elvis Presley (in, for example, the film *Blue Hawaii*) exported the beach culture across the world. In more recent times, the television program 'Baywatch' has kept the excitement of the beach alive – at least in our living rooms. The number of people going to the beach has increased as the human population has grown and middle-class incomes spread. The warm waters and sunny weather of the French Riviera and the Mediterranean (and their rivals) are no longer the sole privilege of the aristocracy. Globally, there is now a large middle class with access to convenient and fast international transport. This cannot be over-emphasised. Without mass transport (originally trains, then, in the twentieth century, steamships followed by the motor car and jet passenger aircraft from the 1950s) tourism would not have grown and continue to grow, as it does today. A Japanese tourist wishing to spend the traditional six nights in Hawaii would have – in days of ship travel – spent the whole holiday travelling!

The Asian middle class

Until very recently the growth in tourism was concentrated in the West and, in particular, the northern hemisphere for obvious reasons (wealth and the relative proximity to the tourist cities of neighbouring countries). By the end of the twentieth century, new groups of tourists had emerged from all corners of the globe. In particular, the fast-growing middle class in Asia is creating hundreds of millions of new tourists. The new rich Chinese are increasingly becoming tourists in the traditional drawcards of London, New York, Paris, Rome and Sydney. It is not just an outward flow, however, as the inquisitive seek to better know the emerging Asian giants and to visit them. Tourism to China is 'exploding', to use a term favoured by the tourism promotion people. With rapid Chinese industrialisation and overseas sales of products comes a desire by those who purchase Chinese goods to see the source of these goods and to explore this nation before it takes on a too-familiar Western face.

THE IMPACTS OF TOURISM

The positives

Wherever tourists gather in large numbers there are impacts – usually positive financial ones for the local hospitality industry. It is for this reason that both poor and rich countries welcome tourists, and many go much further and spend considerable amounts of money promoting their attractions. As a rich country, Australia is an excessive promoter of tourism – movies are made that are little more than advertisements, so-called tourism 'ambassadors' are appointed and formal promotions are extensively used. The financial benefits flow through to those who supply the hotels and restaurants with food, resources and labour. Other suppliers, such as transport companies, obtain financial benefits as well. It is because of the significant flow-on (multiplier) impacts that tourism is such an effective generator of jobs.

The negatives

Tourism can cause negative impacts on the local culture, and more often than not, negative environmental impacts grow with tourist numbers. What tourists are seeking from their experience has changed over time, meaning that the benefits and pressures will be moved from one location to surface at a new one. Too often, environmental damage is done and rehabilitation neglected at the site that acted as the original drawcard for tourists.

TOURISM AS AN EXPORT BUSINESS

A particular issue needs attention before we move on. International tourism is a very valuable export business, obviously different from traditional export industries. You can ship canned fruit, oil, minerals and much else to other countries; however, you are unlikely to meet your buyers face to face. In the case of tourism, you deal with every customer in your own country. International tourism means a personal relationship with your buyer: the tourist. This is both an opportunity and a challenge. You are selling experiences, not shirts or oil or whatever. Shirts have to be manufactured: the cotton grown and silk spun and weaved. Oil has to be found, wells dug and massive machinery put in place. A whole range of resources has to be employed: machinery, fuels, labour, power. Yet a tourism product more often than not already exists. Either nature has made it – if it is a beautiful environment such as a coral reef, an alpine peak, or a rainforest – or previous generations have erected it – if it is a pyramid, cathedral, shrine, mosque or some other cultural or historical place.

Rich in natural resources

Some countries are rich in oil, some in minerals, some in agricultural land and some in water. They use these resources for economic development. Some countries are rich in UNESCO-listed World Heritage properties. These are as good as oil in the ground (in fact, they are better, as they are not used up to ultimately disappear). Properly managed World Heritage properties can be used to earn good income from paying tourists and part of that income can (should) be used to fund the provision of clean water, sanitation, education – the things the poor need for a better sustainable future.

However, it is not all plain sailing for a country that is heavily reliant on tourists for foreign exchange earnings. The problems of tourism have to be understood. It does not take much to turn the tourism tap off. Health scares or threats of terrorism will do just that. Countries that rely on (or hope to rely on) ecotourism – or tourism generally – have to be willing to counter these threats to their economy. Protecting the environment on which tourism depends is the first task – this is taken as given – but after that any perceived threats can, and must, be dealt with.

Not all plain sailing

Safety and security factors impact upon tourists and the economies that rely on tourist income. How safe is international travel? As we will see, the number of fatalities speaks for itself. Our views on most things are, understandably, shaped by the media and, because of media sensationalism, we need empirical data to put stories of the dangers into perspective. The media grabs our attention (and hence sells advertising space) with dramatic stories, and there is little more dramatic or heart-wrenching than a twenty-first-century terrorism attack, a previously unknown disease spread by passenger jets or a tsunami. However, the media does no-one any service by overstating the threats, as it does too often. In the early years of this century, international tourism experienced some major shocks and, after each, tourism was derailed for a period: numbers of travellers fell dramatically to the affected areas, airline earnings plummeted and a sense of foreboding settled on the industry. Terrorism, disease and natural disasters are a major cause of tourism's troubles, but should the traveller be afraid?

Throughout recorded history the world has experienced terrorism. Yet, in the past, terrorism was usually confined to one country, or more likely, to parts of a country, and the aware traveller did not venture to the dangerous areas. Matters have changed. The terrorism we witness today is extremely difficult to comprehend.

Bali – paradise lost

In 2002 a terrorist attack on a major tourist destination, Bali, was the terrorism story of the year. In terms of numbers killed and wounded this was by far the most significant event of the year, and hence got extensive publicity. The targets were a nightclub and a bar frequented by foreigners (in particular, Australians) in a Hindu province of Muslim Indonesia. Fundamentalist Muslims murdered tourists and members of another religion – although some Muslims were also killed – to further a fanatical aim. One can only surmise that the goal of these terrorists was to stir the embers of the ever-flickering clash of cultures and to goad their perceived enemies into retaliating, which, they hope, would lead to more recruits joining the terrorists' ranks. It was a clearly constructed strategy – a tactic as old as human society.

The Bali bombings' impact on tourism, globally and locally, was very significant. Visits to Bali dropped dramatically. While two years later (by the end of 2004) many businesses had returned to normality, some individual resorts still had not managed to get their occupancy rate to a profitable level. Not only had many Balinese people lost their lives, but also the workers in the resorts and their families suffered loss of income. Families were living in one-room flats, husbands were working two jobs and their wives were also working while struggling to look after young babies – these were the hidden losers. Indonesian versus Indonesian in the name of a god!

Not all risks are the same

Let us put death and injury from terrorism attacks into perspective. During the twentieth century we killed worldwide 150 million people in wars, 170 million by government repression,[2] 100 million by famines, 25 million in motor vehicle accidents, 14 million by genocide and 10 million in natural disasters. This is an average of nearly 5 million people killed per year. Except for the natural disasters, we humans carried out these killings. The United Nations has estimated that we managed to kill about 3.6 million people in civil wars during the 1990s alone.[3]

Yet terrorist attacks are designed to kill. For the tourist the possibility of a terrorist attack involves exposure to involuntary risk. If we go skiing or scuba diving we are taking a voluntary risk; if we drive a motor car we are aware of some level of risk and how we assess the level depends on the skill of the driver. For well-trained, defensive drivers the risk would be assessed as extremely low, while it would be much higher if we were in the car with an inexperienced, tired or drunk driver. And, of course, the risk associated with vehicle use differs according to the local driving culture; some cultures tolerate far worse driving behaviour than others. But there is a worldwide downplaying of the fatalities caused by motor vehicles: for example, in Australia where driving behaviour is considered to be good by world standards, 20 times more Australians were killed on the roads in 2002 than those Australians killed in the Bali murders. There was little media or public interest in these road deaths, yet most of the people killed on the roads were as innocent as the Bali victims, and they all have family and friends. My guess is that the media neglect of the road toll influences our attitude to what happens on our roads on a daily basis.

Terrorism must be beaten, but let us not overrate its importance. The relative risks, should not deter tourists from experiencing different cultures and religions, sitting down for a meal with people of different colours, whether it be in Bangkok, Barcelona or Baghdad. The more that we do this, the more advocates for peace and sustainability there will be.

The exotic disease syndrome

Consider another issue that frightens travellers: disease. Various threats emerged in the first years of the twenty-first century. A major one, in the media at least, was severe acute respiratory syndrome (SARS). It first appeared in early 2003 and spread rapidly among staff in Hong Kong and Vietnamese hospitals. Two days after it first appeared, SARS had been taken overseas from these countries, moving rapidly along major international airline routes. Singapore and Toronto reported patients with SARS-like symptoms. Concern was heightened on 15 March 2003 when a SARS outbreak alert was issued after a medical doctor displaying SARS symptoms boarded a plane returning to Singapore from New York. The man was quarantined in Germany. This was Europe's first SARS case. This event triggered a second, stronger, alert that provided advice to international travellers about the symptoms of SARS, and gave the disease its name.

The impact of SARS on international travel in 2003 was significant. With widespread media coverage of the disease, international travel fell in some places by 50 to 70 per cent. Along with this, hotel occupancies in the same countries fell by 60 per cent. This caused numerous businesses, particularly those focused on tourism, to fail.[4] By 7 August 2003, there had been 8423 reported cases of SARS worldwide, and approximately 774 deaths. More than half of these cases were reported in China. Canada and Singapore both reported small but significant outbreaks.[5] Many destinations, especially in Asia and the Pacific, welcomed less than half their usual number of tourists in April and May 2003. This was a minor and short-lived economic disaster.

Travel has always been vulnerable to the reporting of events that are, or can be, portrayed in the media as a threat to personal health, safety and security, and SARS proved to be a classic case of scaremongering.

The real health threats

The big killer worldwide is malaria, with over one million deaths per year. There are approximately 3000 deaths from malaria per day in South Africa alone – per day! Malaria is endemic in the poorest countries in the world with 300 to 500 million cases each year. It is a major contributor to poverty because, if it does not kill, it reduces the productivity of those infected. It has been estimated that the economic loss in Africa due to malaria is in the order of US$12 billion per year.[6] Consider AIDS next: there were three million deaths from AIDS in 2003 and there were 40 million living with the disease.[7] Next take cholera: in 2001, 58 countries reported 184 311 cholera cases and 2728 deaths to the WHO, the vast majority of which (94 per cent) were in Africa.[8]

Tsunami

Another major cause of death are natural disasters and the tsunami in the Indian Ocean on 26 December 2004 was one of the most devastating natural disasters of recent times, although there have been far more lives lost in other natural disasters in the past. This tsunami killed at least 300 000 people, the majority in the province of Aceh in Indonesia. However, it was the fact that the internationally renowned tourism island of Phuket in Thailand was hit that made the tsunami a First World story.

MAKING TOURISM WORK FOR PEOPLE AND THE ENVIRONMENT

From a sustainable development perspective, the task is to work with other peoples' desires and capacity to travel for pleasure so as to ensure that tourism can reach its potential as a force for social and economic good, while not destroying the environment, which is the source of pleasure and fun. That force for good does not just mean the benefits that come from spreading understanding between people of different colours and languages, but also includes the very positive role of tourism in sustaining the world's valuable ecosystems. How can this be when I have just alluded to the threats to the environment that tourism imposes? To elaborate on this it is necessary to turn to what is commonly called 'ecotourism': the form of tourism that promotes the preservation of nature's most precious gifts while providing income and jobs for local (usually poor) communities.

Ecotourism

Nature-based tourism (for example, visiting national parks and World Heritage sites) grows in significance yearly. Data gathered from around the world indicates that natural environmental settings are, at the very least, a crucial backdrop to much tourism. Nature-based tourism is a subset of ecotourism. It is not full-blown ecotourism, as we will see.

Ecotourism developed rapidly from the early 1990s. The first national ecotourism association in the world was formed in Australia in 1991. There are many explanations for the rapid growth of ecotourism, ranging from the psychological (for example, a desire for human closeness to nature), to the influence of popular education programs (for example, the natural history television shows), to the very significant change in attitudes to the environment. Ecotourism has the potential to help protect nature by linking preservation to profit-making. In making this connection between the environment and economics, ecotourism is an example of the application of the fundamental principle of sustainable development: to marry the environment and the economy.

Nevertheless, we should not overstate the case for ecotourism. It is still special-interest tourism (not yet in the mainstream) and there are many kinds of special-interest tourism types competing for the tourist dollar. Ecotourism is, as mentioned, nature-based tourism. It is low-impact tourism. It is, importantly, both enjoyable and a learning experience for tourists. It is tourism that respects local cultures – however, it is not cultural tourism in which the primary focus is on experiencing a different culture. Ecotourism gives something material back to the community in recognition of the satisfaction gained by the tourists; for example, sponsoring education programs for local children. Each of these criteria has to be treated seriously if ecotourism is to be the source of progress towards sustainability. There is a danger that not-so-genuine operators will attempt to cash in on the 'brand name' of ecotourism. This will spoil it for all: the genuine operator, the phoney one and the tourist.

Priceless assets

Ecotourism is linked to the distinctive nature of a place and the valuable assets, such World Heritage properties, that exist in many countries. Contemplate for a moment attractions like these: the Galapagos Islands (in the Pacific), Komodo National Park (Indonesia), Royal Chitwan National Park (Nepal), Serengeti National Park (Tanzania) and the Bwindi Impenetrable National Park (Uganda) – this last one might not suffer from overcrowding!

Ecotourism is not adventure tourism (or extreme tourism) – you can bungee jump anywhere! It is not simply sightseeing, because sightseeing does not require tourists to engage with the place or the people. Firsthand (or some would say 'hands on') experiences such as being able to smell the flowers and the vegetation as it rots, being able to hear songbirds or pull leeches from your skin in a rainforest are typical experiences of ecotourism. Not everyone's cup of tea!

Good interpretation services, cultural sensitivity and involvement with the local community are the three criteria that differentiate ecotourism from its cousins, green tourism and sustainable tourism. Green tourism can be defined as the application of cleaner production, pollution prevention and eco-efficiency methods to tourism (for example, at a simple level, saving water in hotels by reusing towels). Sustainable tourism has a slightly different definition as it has to meet the ideals of sustainable development, but does not have to be focused on nature or present an interpretation of nature to the tourist. A large city hotel could operate according to sustainability criteria without reference to the aesthetic appeal of nature. These are the differences.

Big can be better

There is one issue that causes misunderstandings about ecotourism and needs to be resolved on a case-by-case basis. It is the question of appropriate scale. Some ecotourism proponents advocate small-scale (village-level) operations, as you might find in off-the-beaten-track places in developing countries. The advocates of village-level tourism are unlikely to think of modern villages as fitting ecotourism guidelines, and they certainly would not countenance visits to nature's beauty spots by the large traditional 51-seat tourist buses. The 'small is beautiful' proponents don't believe that large operations can meet ecotourism criteria. Yet scale does not matter if the tourism activity causes only low impacts while still meeting the positive criteria spelt out above. Some large-scale tourism operations can have less impact (because of their design) than the cumulative impact of many small-scale operations in the same locality. Is it not preferable to show hundreds of thousands of visitors each year a rainforest from a cable car strung above the trees than have the same number tramping on foot over the forest floor? In the early 1990s, there were protests (some involving force) against the construction of a cable-car over the canopy of a rainforest in the World-Heritage-listed wet tropics in northern Australia. In this case, the green protestors were targeting the wrong project.

Eco-labelled tourism

One can understand the concerns of supporters of small-scale ecotourism, because until fairly recently – until the advent of accredited or certified ecotourism products – there was a propensity for the large-scale operations to neglect or downplay environmental features and impacts. Now, in a number of countries, genuine ecotourism operations proudly display an eco-label and jealously guard their reputation. The question of scale is likely to be answered on a case-by-case basis, much depending on the type of tourism operation, the transport methods used to bring tourists to the site and to travel around it, the management philosophy of the owner or manager, the government regulations and the type of infrastructure involved. Each and every one of these can have a high or low impact.

The principles of ecotourism

Eight principles of ecotourism have been suggested for tourism operators whose goal is ecotourism. These are based on the world's first certification scheme, the Australian National Ecotourism Accreditation Program. The broad criteria are presented in Box 2. They do not explicitly focus on matters that are dealt with in more specialist literature, such as how to achieve minimal impact while, say, boating by using an electric motor rather than a petrol–diesel one, or how to 'design with nature' (buildings that make best use of natural sunlight, breezes, water resources), or how to accurately interpret nature, or how to undertake an ecotourist waste audit, or how to develop and use an environmental management system. It would take us too far afield to discuss these in any detail. However, the ideas they convey should be obvious.

Box 2:
Ecotourism criteria

Ecotourism:

- focuses on a direct and personal experience of nature
- provides opportunities to experience nature in ways that lead to greater understanding, appreciation and enjoyment
- represents best practice for environmentally sustainable tourism
- contributes to the conservation of natural areas
- provides ongoing contributions to local communities
- is sensitive to different cultures, especially local cultures
- meets customer expectations
- is marketed accurately and leads to realistic expectations.

TRAVEL FOR GOOD

Travel is, or at least can be, a force for good. People with different values and attitudes get to know each other, and people get to better understand the limited size of the globe. If travellers become ecotourists and if the tourism industry promotes and develops ecotourism, a positive step towards sustainable development will have been taken. However, to achieve this there are important things that both governments and the tourism industry must do.

What we might call the mainstream part of the industry – the big hotels in Bangkok, New York or Beijing – have to become 'green'. Simple environmental audits can ascertain the extent to which such hotels are meeting eco-efficiency guidelines with regard to energy use and water and waste management. In most hotels an audit is very likely to find that savings can be made in all these areas.

Using fewer resources is going to mean more profits. Of course, it is not just the accommodation side of mainstream tourism that can move to become eco-efficient; the same applies to transport, eateries, theme parks and, of course, air travel. Making air travel environmentally sustainable will be a very difficult challenge – much will depend on the progress made with alternative fuels, lightweight materials and economies of scale.

Eco-hypocrites?

Some might think that travel, particularly that covering the vast distances that aircraft do, is not consistent with a sustainable lifestyle. This is right and wrong. The available data indicates that aircraft emissions are only a small component of greenhouse gas by the transport sector. However, the fact that emissions occur directly into the upper atmosphere does cause worrying problems. There are, though, far more urgent issues on the ground (so to speak) than that of air travel. After we change from fossil fuels for motor vehicle transport – an urgent job – we can then deal with aircraft.

Ecotourism should become the future of nature-based tourism. Hence, the developing countries can do much more in using their magnificent World Heritage properties as drawcards. In doing so, they should persevere with a policy of dual pricing; that is, charging a considerably higher entry fee for foreign visitors to boost the local economy. After flying from Berlin or Stockholm to South-East Asia, the mid-Pacific and central Africa, tourists will not miss the few dollars entry charge to a World Heritage site. The provision of accommodation, local transport, food and other services associated with tourism should be in local hands. The expertise to establish these businesses can be brought in if necessary, but the profits should not be taken out.

A green logo

The other tasks for ecotourism operations, in both developed and developing countries, is to put in place a universal third-party accredited and monitored certification scheme – a 'green logo'. This system is in place in a number of countries, and has been for some years. If based on the ecotourism criteria described earlier, this will not only attract tourists but also guarantee environmental standards, and keep competitors honest. To advertise as an ecotourism business, you have to meet the appropriate standards. Do this on a large scale and we will have a prime example of how the environment and economics can link, and the dual benefits that can result therewith.

1   Tony Perrottet, *Route 66 AD: On the Trail of Ancient Roman Tourists*, Vintage Australia: Sydney, NSW, p. 6.

2   United Nations Development Programme (UNDP), Human Development Report, 2002.

3   United Nations Development Programme (UNDP), Human Development Report, 2002, p. 2.

4   World Health Organization, 2003.

5   World Health Organization, 2003.

6   World Health Organization, 2001.

7   World Health Organization, 2003.

8   World Health Organization, 2002.

5

DROPS IN THE OCEAN: FIJI AND FRIENDS

Bula!

The universal Fijian greeting.

Here we are concerned with the major problems retarding – and seriously so – progress towards sustainable development in a set of particular circumstances. In no order of priority they are small land area and small population (bringing their own costs – diseconomies of small scale in particular); distance from major markets (imposing high costs for both imports and exports); ethnic conflict; open access to the commons (see Chapter 2 for information on the tragedy of the commons); and the lack of secure property rights. While we can find these problems in various parts of the world, their overriding effect is most evident in the vast South Pacific. This is an area of tiny nations, far removed from major markets (Australia is the only close economic power), with relatively little influence on their own future. Global warming, the most serious tragedy of the commons the world faces, is predicted to increase sea levels to the extent that the low-lying coral atoll nations could disappear.

However, the situation faced by the South Pacific island states is not all gloom. Ecotourism is an industry with much promise in this sun-kissed, generally relaxed and friendly part of the globe. There is also scope for most of these tiny nation states, with their extensive exclusive economic zones, to gain considerable economic (and employment) benefits from their fisheries' resources.

Countries small in area and population, particularly if they are isolated, face numerous barriers to development. Small countries in Europe (for example, Switzerland) or Asia (Singapore) or the Middle East (United Arab Emirates) do not face the costs of isolation, in particular the transport costs involved in both exporting and importing goods. Yet an isolated country can look to its natural and cultural attributes as the base on which to give its people a good, sustainable life.

THE CASE OF FIJI

The country of Fiji is the giant among the pygmy island states of Oceania.[1] If Fiji cannot survive in the twenty-first century, what hope is there for the much smaller, more fragile and economically dependent island states of the Pacific? Fiji is a three-hour flight from Sydney, a five-hour flight from the US state of Hawaii and about a seven-hour flight from Tokyo. These relatively short flight times highlight the tourism potential of Fiji.

The landmass of Fiji is almost 19 000 square kilometres, yet its hundreds (thousands, if every single rock or coral outcrop is counted) of islands are scattered over 1.3 million square kilometres of the South Pacific Ocean. The country's population is approximately 900 000 people. Population is the most obvious indicator of size and importance in a region where nation-states are miniscule in both size and population. The smallest country in Oceania is Nauru, covering just 21 square kilometres; countries with the smallest populations in Oceania are Tuvalu and Nauru with approximately 11 000 people. Both China and India have populations well over one billion people.

The lustful, leisurely life

Common images of the South Pacific of a leisurely life tell only a small part of the story of the peoples of the Pacific, and their trials over millennia. The South Pacific Ocean covers 28 per cent of the world's surface area. It focuses the mind if we say this is about one-third. One third! It holds more than 50 per cent of the globe's free water. Yet for most people of Europe, the Americas, Asia and Africa, the South Pacific may as well not exist. Not much happens there. There is little to fight over.

A generation ago the French had a sure-fire way of drawing attention to the Pacific – testing atomic bombs. Today it takes a coup or rebellion to make any impression on the world stage, and the coups are becoming commonplace in recent years – and that diminishes their newsworthiness. Even rebels taking the whole Fijian parliament hostage in 2000 had little impact other than in neighbouring Australia and New Zealand. In fact, it took the first Australian military death since the Vietnam War to get the South Pacific on the front pages of Australian newspapers.[2]

A middle-ranking country

Today, Fiji is a middle-ranking developing country, coming in at 92 out of 177 countries, or about halfway, on the 2005 human development index (HDI). Life expectancy is close to 70 years and adult literacy is around 93 per cent. These are impressive numbers. Per capita gross domestic product (GDP), measured in purchasing power parity (PPP) terms, was US$6049 in 2005. However promising these figures, they represent a slip in status for Fiji since 2003 when it ranked 81 on the HDI, had a higher life expectancy and higher real income.

Notwithstanding the setbacks, Fiji has managed over a 25-year period to improve from 0.654 to 0.762 its HDI score. These numbers mean little to most of us, so let me put them into perspective. Fiji's neighbour, the Solomon Islands (one of the poorer countries in the world), had in 2001 a HDI score of 0.602 and came in at 129 out of 177 countries. Fiji was ranked higher than this a generation ago. In the early 1970s, life expectancy for Fijians was approximately nine years lower than it is today. Clearly, Fiji is not a poor country in comparison to its poor neighbour; however, it is still some considerable distance from reaching its potential HDI ranking, which is much higher than where it sits today.

Nature under threat

Fiji's natural environment is in relatively good shape, yet population pressure and the willingness to exploit natural resources without thinking for the long term suggest that this won't be the case forever. Of a total land area of 1 827 000 hectares, tropical forests account for nearly one-third, but only one per cent of these forests are protected. The small amount of mangroves that still exists (about 50 000 hectares) is completely unprotected. As for animal life, four species of bats are under threat, so are some reptiles and twelve bird species.

In the postcolonial period,[3] the lack of exploitable resources in most of the countries of the Pacific region has meant that the major global powers have not shown much interest in them. The only valuable resources of significance are the pelagic fisheries, tuna being the most profitable. Some of the tiniest countries in land mass have, due to the highly scattered nature of the islands that comprise each nation, enormous exclusive economic zones. Kiribati is the prime example. Large territorial zones give some countries power to make some income from fish caught in their waters by foreigners. The income is meagre, as is the bargaining power of these Pacific Ocean countries, who are at the mercy of the large fishing fleets from Japan, the USA and Taiwan. The financial return from access fees and licences paid by foreign fishing companies is a tiny 5 per cent of the US$2 billion the South Pacific harvest brings on the world's retail seafood market. If that is not dismal enough, much potential income is lost by fish theft, which is sometimes given the politically correct name of pirate fishing. Call it what you like, it is robbing the poor. Admittedly some of the robbers are also poor, but be mindful that they are the hired hands of wealthy Fagans.

Fish theft is costing the Pacific nations somewhere between US$100 to $400 million per year, more than they earn in fees from foreign fishing companies. The modern pirates are not the swashbuckling heroes or rogues flying the skull and crossbones, rather they sail under flags-of-convenience bought on the internet for a few hundred dollars. Who knows who owns the vessels, pays the crews and reaps the profits from their robbery? The thieves are not on the high seas, but in an ocean that belongs to some Third World country.

The Icelandic solution for protecting marine resources is not available to the Pacific Nations. They do not have coastguards capable of patrolling the vast areas they own, let alone engaging in 'tuna wars' of equal resolve to the North Sea cod wars.

Innocence lost

It is a cliché to suggest that the Pacific lost its innocence when the European explorers, traders, missionaries, beachcombers, artists, layabouts and colonisers arrived. But then, much depends on what we consider innocence.

Still, the notion of a loss of innocence reappears throughout recent Fijian history. The 1987 coup in Fiji by Colonel Rabuka, following the election of the Indo-Fijian-'dominated' coalition government[4] on 12 April that year, is today described as 'the day Fiji lost its innocence'.[5] This, of course, is a different form of innocence to that lost when the Europeans arrived. Postcolonial Fiji, with its democratic institutions given to it by the UK, was supposed to be another New Zealand, Australia or Canada, only on a smaller scale. Military coups were not meant to be in the model.

Why the coups?

It is worth thinking about the underlying causes of the Fijian coups. Were the coups about race, religion (Rabuka said at the time that 'God gave us this land' – 'us' being the ethnic Fijians, not the Indo-Fijians), social class, economic class or simply the economic self-interest of certain unnamed (unknown) individuals? Were religion and race used by powerful economic interests to gain support for insurrection by the generally tolerant Fijian population? These possibilities are the ones much discussed in what has become a substantial literature on the coups. The real story is yet to be told, but ever so slowly, more and more relevant information is being uncovered. The coup plotters are being outed by their foot soldiers, underlings who were captured and jailed while the masterminds remain free to enjoy whatever financial benefits they gained by the overthrow of the government. At the time of writing, the guilty have not been apprehended or charged.

A little history helps

While the possibility remains for self-interested entrepreneurs and politicians to stir ethnic tensions, the Fijian people will not have a sustainable society. Draw a line in the sands of history somewhere in the distant past and this will have given some people cause to fight their neighbour. The Fijian history is a classic case – the line was drawn by ethnic Fijians the day before the first Indians arrived. You have to go back further in history to understand why this line and not another, as Fiji has a history of invasions and colonisation.

Today's indigenous Fijians are a Polynesian–Melanesian mix, although they are often described in the popular press as Melanesians. Over millennia, with the end of the major migrations and the establishment of permanent settlements and distinct family and tribal groups, different lifestyles (or cultures) became established on different islands (even on different parts of islands). Notwithstanding a common (mixed) ethnicity, tribalism replaced a common culture. Each tribe could find a reason or reasons to distinguish it from its neighbouring tribes, and conflict in the form of tribal wars over land and its resources became a normal way of life. Until the twentieth century, the South Pacific history was not one of idyllic peace (as advocates of the noble savage would have us believe) but one of continuing tribal wars. How do we explain these wars?

Today, people who call themselves ethnic Fijian will, if stirred by the likes of Rabuka, assert rights over the near half of the population who just happen to be born in Fiji, but from ancestors who arrived only 100 years ago. In the Fijian scheme of settlement, invasion and peaceful immigration, the Indo-Fijians are the most recent arrivals. Throughout history there were successive waves of invaders to the Fijian islands, but unlike the earliest settlers – whether the ancient Polynesian or Melanesians or the British in the 1800s – the Indians came as workers and helped to build the Fijian economy. They came peacefully, yet they are subject to a desperate racism today.

Settling the South Pacific

The 'where from', 'when' and 'how' of the first Pacific migrations and settlements are not unanimously agreed; however, something approaching a general consensus has emerged through scholarly debate, the finds of archaeology and the anthropological adventures of the Norwegian Thor Heyerdahl in his Kon-Tiki expedition in 1947. That Heyerdahl's theory of a migration from South America proved wrong is not the point. He stirred the quest to uncover the ancestry of the people of the South Seas and showed that it was possible to travel in the most primitive raft from South America to the Pacific islands. The more recent research of other anthropologists and archaeologists gives us an understanding of how distinct racial types have come together in modern Melanesians and Polynesians and, together, these races form the background of modern ethnic Fijians. The following story is the one commonly told.

A few find the Pacific

People first came to Oceania (which includes Australia) about 40 000 to 60 000 years ago. Sea levels were low due to an ice age and Australia and New Guinea were joined. The first people to reach Oceania originally started their migration from South China. They moved into South-East Asia and then some settled in New Guinea, some in Australia, and others ventured far into the South Pacific Ocean.

On the basis of this historical construction, the peoples of Sumatra, Java, Molucca, New Guinea and the Aboriginal peoples of Australia have the same ancestors. These first people – those who get to be known as 'First Nation' peoples – were conquered or were pushed into the interior of their lands by the next round of arrivals. As in most, if not all, territorial invasions the new and old peoples intermarried and formed 'new' cultures and races.

The contemporary cultures of the Pacific islands are not those of the very first people to arrive there, but rather are cultures that formed over millennia as invaders, some originating in the same Asian homelands as the very first settlers, mixed with the early inhabitants. This mixing of cultures from waves of sequential invasion is the case the world over – humans, if they are willing to recognise it, are a cosmopolitan lot.

The second phase of invasion and settlement in the South Pacific occurred between 30 000 to 40 000 years ago. These new people took control of coastal areas with their abundant marine and terrestrial resources. New arrivals moved into the New Guinea islands and then the Solomon Islands. They mixed with the first phase of settlers and with subsequent arrivals to become what we today call Melanesians.

Melanesians and Polynesians settle Fiji

In the long sweep of history, Fiji has been settled by various peoples from South-East Asia, then invaded by the people who now call themselves ethnic Fijians. These 'original' Fijians are a mix of two different cultures, the Melanesians and Polynesians.

The first settlers in Fiji were Polynesian. They were typically light-skinned, broad-nosed Austronesians. Like the Melanesians, they also came from South China or South-East Asia. They 'leapfrogged' the islands already occupied by the Melanesians. Not only did these people settle Fiji, but also Samoa, Tonga, the Society Islands, the Marquesas and the distant islands of the South Pacific. The Polynesian migration was completed in the period 1500 to 1000 BC.

The third migration phase into the South Pacific brought Lapita pottery[6] to Melanesia. Discoveries of it in Fiji have been dated at 1290 BC, but this is well after it is believed Fiji was first settled. The eastern Solomon Islands are thought to have been the departure point for the voyages by these excellent pottery makers into the rest of the Pacific.

A common ethnicity

Fiji's early settlement did not end there. The Melanesians, who had been leapfrogged by the Polynesians on their journey to Fiji, arrived in Fiji about 500 BC. Once again a mixing of people took place, and yet another new 'race' was created, the one we know today as ethnic Fijians. This historical sketch illustrates that much that is spoken and written about race and ethnicity today is nonsense. Those who seek so-called purity of race have no understanding of our cosmopolitan past. That does not stop the ideologues – and won't stop them in the future, unless our children are made aware of the mixed ethnic history of most peoples (whether in the United Kingdom or Fiji).

The Europeans arrive

The first European we know to have sighted Fiji was Abel Tasman in 1643. Like many others in that era, Tasman was searching for the place we now call Australia (then Terra Australis Incognita). More than a century later, in 1774, James Cook sighted Fiji. However, it was not until the mutiny against William Bligh in 1789, and his lucky escape that took him through what is now known as Bligh Waters, that Europeans started to get an accurate picture of the country of Fiji.

By 1800, tales of abundant stands of sandalwood enticed traders to the Fiji islands. The wood was valued for making joss sticks and incense, valuable products then, if not now – except in hippie circles and Asian communities. Within a decade and a half, the forests had dwindled and the industry collapsed.[7] Next came the traders in bêche-de-mer (large sea slugs that, when dried, are prized as an aphrodisiac in parts of Asia). Shore-based processing facilities were established where bêche-de-mer was smoked, dried and shipped to China, and a European community based on this enterprise was established. Next came the 'beachcombers', an improbable mix of fortune-seekers, layabouts and escapees from civilisation (such as it was in the 1800s). These were the nineteenth-century version of rich hippies – rich because they had to pay for their travel aboard merchant traders. By the 1830s, a beachcomber settlement existed at Levuka. Then, as whaling came to the Pacific, crews called into Fiji for supplies, some fun and some rum. This added to economic development in what had already become a rather wild South Seas outpost. European trade was one thing, but unscrupulous traders cheated the local people. Local chiefs, who were willing to do deals with the foreigners, became extremely powerful and rich.

Converting the pagans to Christianity

The coming of the Christian missionaries was a great shock to Fijian culture, which was to change forever. The first to convert were the pagan chiefs. Their underlings followed, without question – if it was good enough for the chief, it must be for them. They were also aware of the chief's power.

Before becoming Christians, the Fijians were fierce cannibals, worshipping a whole range of gods. The missionaries were mainly Methodists. They put the Fijian language into script and the women into long dresses. The former was a major step in the transition of Fiji into a modern state. With regard to the latter, in the twenty-first century, Fijian women, wearing the same type of 'Mary dress' as their great-grandmothers, must wonder why female tourists wear next to nothing.

Cotton farmers come then go

In the 1860s, cotton farmers arrived in Fiji in large numbers. During the American Civil War the massive disruption to everyday American life (particularly in the South) meant that the supply of cotton for European mills fell dramatically. Various places around the globe with the right climate, soils and labour force were tried as alternatives – the 'right' labour force being one paid as little as possible. Fiji picked itself as cotton country. When the Civil War ended, so did most of these substitute cotton-farming ventures. Once the American South began growing cotton again, the industry folded in Fiji and in other countries, such as Australia, that had been wartime producers.

Religion, ritual and revelry

In the era we are discussing, Fiji must have had an interesting mix of religion, ritual and revelry. It was not a country in any formal sense, rather a piece of paradise torn this way and that by rogues, vagabonds, religious zealots, entrepreneurs and local tribal chiefs. The first attempt to form a quasi-federal government was made in 1865 when a group of chiefs from the major islands got together with the objective of bringing their collective power to bear against the lawlessness that had come to typify Fijian society. This experiment of cooperation failed after two years of endeavour.

The big chief and Britain

The next political initiative was an attempt to form regional governments. This idea, likewise, was not successful. But perseverance was to pay off. In 1871 Ratu Seru Cakobau, an extremely powerful chief (some chiefs were extremely powerful, as a consequence of land ownership and their fierce, loyal fighters), formed a central government in partnership with a foreign planter, John Thurston. This pattern of chiefly and British interests joining together for mutual benefit continued right throughout the colonial period. This coincidence of interests was enshrined in the Fijian constitution and can rightly be viewed today as a politico-cultural impediment to Fiji becoming both a genuinely democratic and an economically progressive country.

In 1854, Cakobau had become a Christian. Not only did he have great authority over the local people, but also he now could claim to have one god, deemed to be more powerful than the array of Fijian gods, on his side. It was an easy step for him to appeal to the British Christian king for protection. In a deal with King George of England, Britain became the de facto guardian of Fiji and, even if they didn't know it or agree to it, the Fijian people became Christians in line with their self-proclaimed chief's conversion.

In 1862, Britain had the opportunity to annex Fiji but declined. But its hand was soon to be forced. In 1867, a US warship threatened to shell the town of Levuka over a disputed land deal.[8] British settlers became very concerned and feared for their life and possessions. It is in these circumstances that we should understand the reasons for the formation of a central government in 1871. However, the route to this was not without obstacles. Cakobau did not speak for all Fijian chiefs and there was a rival grouping of chiefs based in the east of the country. And in this situation of disputed power, disorder remained a problem. To protect his position as much as to bring order to the country, Cakobau ceded Fiji to Great Britain and, on 10 October 1874, Fiji became a British colony. From then on there would be no doubt as to who was the supreme law-making authority.

The first British governors had a far-reaching and dramatic impact on the way the new nation developed. Sir Arthur Gordon was the first governor, Thurston the second. Rather than impose direct rule, which would have required a strong police force, an army and a powerful bureaucracy, they opted to rule indirectly, through the Fijian chiefs. This was to cement the role of the chiefs into the form of governance that still remains in place today.[9]

The chiefly system and ownership of land

The chiefly system was based on each chief having total power over 'his' communal lands. The British governors realised that if the nexus between land and chief was broken, their surrogate administrators would lose power, status and authority, and hence be of no real use to Britain. A consequence of the British policy was that the British forbade the sale of native land. Here we have the origins of the economic problem of the twenty-first century: the lack of security over land experienced by common people, including Indians who are Fijian citizens.

The chiefly system was not completely a quasi-feudal regime. Land could be leased and by this means foreigners were able to put it to economic use. But in modern times the restricted nature of the leases are a serious impediment to the development of sustainable farming, a subject we shall return to. The British governors furthermore determined that Fijians could not work on European-owned plantations. The reason was so as to not disturb the traditional rural village culture of the local people. The British took the strategic view that the less they modified the tribal culture the more likely that their de facto partnership with the chiefs would hold strong. If the way of life was not seen to change, the ordinary people would not query the rule of their tribal chiefs or the existence of colonialists in their midst. All very Machiavellian, but should we expect anything different?

The labour shortage

There was a downside to the British policies – they created a labour shortage for all but village work. To solve this, workers from the neighbouring Solomon Islands, Vanuatu (then the New Hebrides) and Kiribati (then the Gilbert Islands) were brought under force to Fiji. When this practice was outlawed in 1875 by the UK *Pacific Islanders Protection Act* there was shortage of labour once again. Sugar had come to replace cotton as the major crop and canecutters were needed in large numbers on the plantations. Governor Gordon had experience in how to solve this particular problem, as he had been a colonial administrator in Trinidad and Mauritius and had first-hand knowledge of the work ethic of Indian sugar-cane labourers. The labour shortage solution was staring him in the face.

The Indians arrive

The first indentured Indians arrived in Fiji in 1879 and continued coming until 1916. By then there were 60 000 Indians in Fiji out of a total population of 157 000 people. Today the Fijian population is approximately 900 000, of which 51 per cent are ethnic Fijian,[10] 44 per cent Indian, and the remaining 5 per cent European, Chinese and people from the other Pacific islands.

The Indians signed a contract to cut cane for five years, after which, for the next five years, they were allowed to lease small farm plots. More than half of the Indians who came to Fiji remained as free settlers after their ten-year contract expired. The nature of this initial indenture system has led to the unsustainable situation today in that Indians lease from ethnic Fijians very small, uneconomic farms.

That this arrangement of leasing small plots was counterproductive is easy to prove in strict economic terms. The economies of scale that are possible with large sugar farms were never going to be realised in Fiji. Furthermore, the Fijian arrangement also had serious social and class impacts, but these are not as easy to document. Many of the ethnic Fijians remained subsistence farmers (and landlords) while an Indian entrepreneurial farming class developed, although most were to remain poor. However, some came to a position where they could exert considerable influence on the country's economy. The more successful Indo-Fijian farmers were able to educate their children and some eventually became city-dwelling middle-class professionals. Today the professional and business class in the major cities is dominated by Indo-Fijians. This itself is a source of envy and is a reason for political conflict.

This economic dynamic was bound to produce problems between the materially successful Indo-Fijians and the less-successful ethnic Fijians. There is a conventional wisdom (expressed by a great variety of experts on Fiji) that ethnic Fijians do not seek material success to the same degree as Indo-Fijians. While this is true of some, particularly those ethnic Fijians satisfied with rural village life, it is not a universal phenomenon. There would not be the ethnic and class jealousy (so easily whipped up) if ethnic Fijians did not care about economic success as much as the Indo-Fijians.

History matters, or does it?

A new ethnicity underpins the most serious economic and political problems in Fiji today. The Fijian constitution mandates a racially based electoral system, categorising voters as 'Fijians', 'Indians' and 'voters who are otherwise'. The fact that there is virtually no intermarriage between ethnic Fijians and Indo-Fijians means that cultural understandings are not deep. This is the case even though both groups work side by side in shops, factories and tourist resorts in the cities and towns. And they are friendly, not conscious of an ethnic divide.

History tells a sad story. It shows us that peace is usually only achieved in lands divided by competing cultures when those cultures come to accept their differences (which often are not significant) and intermingle to become more alike than different. Intermarriage is the greatest force for the achievement of understanding. If you need evidence of how this works, consider the long history of the people who now call themselves British. Their DNA can be traced to the far reaches of the globe.

The Pacific is not peaceful

Fiji is not the only place in the Pacific to struggle with such issues. From the late 1990s, the Solomon Islands has been racked with ethnic strife and a resultant loss of life. Similarly, economics and ethnicity were at the core of the civil strife – call it war – that bedevilled New Caledonia in the 1980s. The Pacific is not as peaceful as the utopian dreamers would have it.

Clan loyalty

A particular characteristic of Pacific culture should be noted. It is important because it works against the common assumption of democracy, in particular the notion that elected governments have the right to write laws that all should obey. In the Pacific, the custom of gift-giving and paybacks, and loyalty to one's family and clan – while admirable at one level – can too easily result in conflict of interest when elected parliamentarians or senior public servants put clan before civil society. The rule of law comes second to obligations to clan groups. Tender bids for major projects go to family and friends. Government houses are sold cheaply to people of the same tribe, while others pay much more. Of course, corruption occurs in many places in the world but in the Pacific it is unwittingly assisted by an old-fashioned clan loyalty. The goal of a greater loyalty (to all citizens) would be applauded, and would help make the modern Pacific a better place. Tribalism is the enemy of a sustainable world society – a point not recognised by Western postmodernists.

FIJI'S MODERN ECONOMY

Today, Fiji's economy is based on tourism, garment manufacturing, sugar production, gold mining,[11] timber production, commercial fishing and processing of dalo[12] and coconut products. There was a lucrative banana industry in the past but it was wiped out by the sigatoka virus. Until recently, kava was exported for use in pharmaceuticals; however, reports that suggested it was detrimental to health devastated the export industry. The industry has suffered further through the spread of kava blight. Subsistence agriculture still remains important to indigenous Fijians with manioc, taro, yams, sweet potatoes and corn being grown.

Exporting water

Natural artesian water is a surprising and well-performing export, with demand from the USA being strong. Fijian bottled water also sells well in Australia. Yet, in a Fijian tourist resort you are more likely to be offered New Zealand bottled water than the local product. This is very difficult to explain, as Australian tourists who make up the majority of visitors to Fiji have shown their willingness to purchase the Fijian product in Australia. Not everything is rational in the world of consumer behaviour.

The trees fall

Fiji produces significant amounts of both hardwood and softwood timber. It is thought to possess the largest green mahogany forest in the world. Currently, exports are worth F$50 million (US$35 million) a year in sawed lumber and woodchips. However, Fiji's forests are not necessarily managed to produce maximum value over the long term. Large tracts of pristine rainforests are being lost each year.

Rainforests are extremely valuable repositories of species biodiversity and are being reduced worldwide on a daily basis. Fiji, like other developing countries, faces the dilemma that good money is paid for wood products while no-one is offering money to protect biodiversity in forests. There is a potential solution. Australia has proven, through its decision to have most of its rainforests listed as World Heritage properties, that ecotourism is a more valuable use of rainforests. The Fijian government needs to consider this alternative use of its forests. Tourism is already one of its two major industries, but whether rainforest tourism can become a drawcard to Fiji is an open question.

Fishing for profits

There is an established fishing industry in Fiji and it is the fifth largest export industry in the county. The government subsidises a tuna cannery that produces 15 000 metric tones of canned skipjack and albacore tuna per annum. Fiji also exports 3000 tons of chilled yellow-fin tuna to Hawaii and Japan to serve their markets. Demersal fish are sold on the domestic market, as fish is a popular source of protein in the Fijian diet. Much more could be done to make both the local and export fisheries economically and environmentally sustainable. It would just take the will to make something better of the country's fisheries. If nothing else, tourists will pay good money for the tropical water delicacies that abound.

Local manufacturing

Fiji's largest export manufacturing industry is garment production. In the last decade or two up to 100 companies have been exporting to Australia and New Zealand. The industry employs 16 000 people. Australia and New Zealand allow products with 50 per cent Fijian content to enter partly duty-free and quota-free. Other tax-free exports include shoes, food, toys and furniture. Of course, these artificial competitive advantages will be eroded and will presumably disappear with the further freeing up of world trade. Sugar will no longer be subsidised by 2010. How long the other export industries are permitted to escape the strictures of competition is unknown, but surely it will not be that far into the future. Fiji's only other major export product is gold.

A TROPICAL TOURIST PARADISE

The agricultural, manufacturing and mining products pale into insignificance when compared to tourism. Tourism has been Fiji's biggest moneymaker since 1989. The industry is worth over F$500 million (US$335 million) each year. It contributes between 16 and 22 per cent of Fiji's gross domestic product. As most of the hotels and resorts are foreign-owned, it makes it difficult to calculate the net value of tourism to the Fijian economy. Gross receipts are often misleading, as 56 cents in every dollar are repatriated overseas to foreign investors or are used to pay for tourism-related imports. Sugar is actually more profitable than tourism for Fiji, as it does not suffer this strong leakage of money, but as sugar is subsidised by the government, it is hard to make simple apples to apples comparisons.[13]

The profit leakage problem

A classic example of the profit leakage problem facing tourism is explored in Box 1. Objects that a tourist is likely to purchase in Fiji for either immediate consumption or as a gift to take home are listed. The first column in the box contains a list of typical tourism products; the second shows the retail price if these items are imported; the third column shows what similar local products cost to buy. While tourism operators continue to purchase overseas, at significantly higher cost, products that they could source locally, they will not maximise their profits. It is tempting to explain this situation by asserting that the quality of the Fijian-produced goods is inferior to the imported substitute, but this is not the case. What belies that reasoning is the successful export of Fijian water.

Box 1: Leakages – souvenirs and sustenance

IMPORTED ITEM: COUNTRY OF ORIGIN	COST IN F$ OF IMPORTED PRODUCT	EQUIVALENT COST IN F$ OF FIJIAN PRODUCT
bottle of water (NZ)	2.00	1.19
bottle of water (Malaysia)	2.00	1.19
bottle opener (Korea)	2.00	N/A
souvenir cap (NZ)	30.00	17.85
beer holder (NZ)	7.50	4.50
nip glasses (Europe)	20.00	N/A
postcard (NZ)	1.00	N/A
coaster (NZ)	10.00	6.00

The dollar leakage problem is one the Fijian tourist industry has to solve. Alert business types would see this as an opportunity. Of course, it is not just the purchase of simple items that keep the benefits of tourism low for the local people. Foreign investors are able to purchase the larger resorts because there is a lack of Fijian financial capital. This has to be recognised as an inevitable outcome of a global economy in which restrictions on the inflow of financial capital to a country are not controlled or curtailed.

Ecotourism

For some time Fiji has been making the most of its natural beauty, climate and easygoing lifestyle to earn tourist dollars. It is in the right time zone for time-conscious Japanese holidaymakers, who can be in the Pacific Ocean or the resort pool by mid morning, after finishing work the evening before. It is similar for Australian and New Zealand tourists – they can get there more quickly and pay less than an equivalent experience in their own country would cost. However, Fiji could do much to limit the leakage of tourism receipts out of the country.

Productivity in tourism could be enhanced with environmentally friendly design and fitting out of hotels and resorts. A viable ecotourism industry, as a subset of the general tourist industry, is there for development and is a means for Fiji to obtain the greatest monetary benefits from what nature has freely given it. The Fijian tourist industry sells the traditional products of the South Seas: sand, sun and (hints of) an exotic culture. This is nature-based tourism with a splash of cultural tourism. While a local ecotourism association exists in Fiji, it needs to make more of an inroad into the mainstream industry if any benefits are to be realised. Yet tourism, or ecotourism, is unlikely to be the only long-term answer to Fiji's sustainable development.

The cost of fuel

As noted previously, manufacturing has developed as an export industry and Fiji exports sugar, garments, gold, timber, fish, molasses, coconut oil, water and 'tourist experiences'.[14] The important imports are manufactured goods, machinery and transport equipment, petroleum products, food and chemicals. Fiji, as do all the small island nations in the Pacific, has to pay a considerable price for fuel (petrol and diesel) to power its domestic transport fleet.[15] Hydroelectricity plays an important role, and solar and wind power play minor roles in Fiji. In the future, geothermal power could become a significant energy source, but both hydropower[16] and geothermal power need further development, not just for economic reasons but also for environmental ones. In the tourist industry, more can be done to reduce energy use by 'design for nature' architecture and the use of solar power, windmills and hybrid systems. If there is an alternative use for the sugar cane, or the sugar-cane lands, it could be ethanol production.

SUSTAINABLE DEVELOPMENT

Let us consider how and where Fijians can build on their already existing competitiveness. Some important questions for Fiji are: Why continue to import rice from Australia? Why import a variety of tourist products, some of which we identified before, that can be produced at lower cost in Fiji? Why not produce horticultural and vegetable crops with much higher returns per hectare than sugar cane?

Fijian farmers could consider the model of the tropical sugar-cane growing area on the east coast of northern Queensland. In a period of 20 years this area has become a low-cost producer of tropical fruits – to the extent that Australian-grown avocados (among other fruits) are exported to South-East Asia. The climate and soils of Fiji are very similar to those of tropical Queensland. There are no convincing arguments against the idea that much Fijian sugar-cane land could be converted to tropical fruit growing.

Clothing and footwear manufacture has grown through preferential access to Australian and New Zealand markets. It is, however, a relatively small industry and its long-term prospects must be questioned as the huge producers in China (and, in the future, in even lower-cost Asian countries) come to dominate world markets.

The potential to expand seafood and horticulture exists, but without a radical change in land-tenure arrangements a horticulture industry will not even commence, let alone grow. This is not necessarily a problem if the land remains owned by the government on behalf of the people – a normal situation in modern capitalist economies – however, any lease arrangements must give a level of certainty in the long term, as leasehold farmers in other cultures have. Fish farming offshore (where small tunas, for example, are caught and grown out in cages) has potential. Australia, for example, has been able to make this a very profitable business. All of these suggestions for the sustainable development of Fiji do not necessarily require foreign investment funds,[17] and all profits and other earnings will remain in Fiji.

Production of bio-fuels

The potential for the development of bio-fuels would make a major difference to Fiji's balance of payments. With a relatively small investment – likely to be funded by foreign aid – Fiji could convert from being a sugar producer to an ethanol producer. The reduction in greenhouse gases would be a real benefit – although insignificant in world terms – and powerful as a symbol to the rest of the world.

OTHER SOUTH PACIFIC COUNTRIES

I should conclude with a general comment about the policy choices of other South Pacific countries as it is not only Fiji that experiences coups that frustrate the formation of a sustainable society.

The Solomon Islands

The Solomon Islands government was overthrown in a coup on 5 June 2000.[18] An appeal to ethnicity cloaked the real reason – command over resources – that was behind this coup. Many of the foot soldiers were unemployed youths or petty criminals with no tribal or ideological basis for their involvement.

Nauru

Another South Pacific country that has suffered recently from extremely poor governance is Nauru. Thirty years ago, Nauru had one of the highest per capita incomes of the world. From phosphate sales each Nauruan (including children) was worth about Australian $500 000.[19] Helen Hughes has calculated that had the profits been invested wisely, each Nauruan family would have had assets worth Australian $4 million in 2004. However, the reality is that the country is 'on the verge of insolvency', while a few Nauruans have amassed private fortunes.[20] It is difficult to ascertain precisely what went wrong in Nauru as there is not enough detailed documentation available. However, poor financial management was the root cause. Had there been a wide-enough spread of investments of phosphate profits the outcome would have been very different.

What to do about Nauru given its dire position? Here I cannot be optimistic. It is the world's smallest republic, a nation of 11 000 people, and with its natural resource (phosphate) depleted and only 21 square kilometres of land it is too small to sustain its population. On various occasions in the past, it has been proposed that the entire Nauruan population be resettled in Australia. It may come to that – but we can imagine a degree of resistance from the xenophobic types in Australia.

Tribal lands

We have focused on Fiji because of its relatively advanced status, but the general picture in the South Pacific is that most of the resources are not being utilised to their optimal level. This means that, with appropriate policies, economic growth is both possible and, given the low level of development, desirable. Most of the countries have made little progress since independence and the lack of secure private-property rights is a major institutional impediment to sustainable development. Unfortunately, a change of land tenure is not on the horizon but perhaps it does not have to be. The tribal land-ownership system that exists throughout the South Pacific could be made compatible with an economically efficient and environmentally sustainable use of land, even though it is not at present. Those who would retain the existing system should be asked to show that it can produce a good, sustainable society.

In Fiji, the single most important issue is to give tenant farmers sufficient incentive to look after land and make long-term investment decisions. Only the national government can make this change, and that is only likely to happen when the people who have powerful economic investment believe it is in their best interest to do so. There has been little indication of this to date. However, economic, social and environmental imperatives (such as the loss of sugar subsidies and the rise of sea levels) will focus the minds of South Pacific peoples and significant changes are inevitable in the next few decades.

A fossil-fuel-free Pacific

As changes in land ownership and the rights to use land are some of the most significant a society can undertake – often resulting in bloodshed – we can expect to wait a considerable time before anything at all radical takes place in this regard. There is far greater promise in changing the South Pacific to a fossil-fuel-free zone through converting sugar-cane production from sugar to ethanol. This does not require any significant change in land tenure. It does require investment in the appropriate technology – which only the government can do – and a willingness to forego taxes collected on imported diesel and petrol. These may seem to be impediments, but on rational analysis of the economics they are not. Believe me, I've done the sums.

1   There are three large island states in Oceania: Australia, New Zealand and Papua New Guinea.

2   This unfortunate person, a peacekeeper, was killed on 22 December 2004 in the capital of the Solomon Islands, Honiara.

3   Only two countries retain colonies in the South Pacific: France continues to assert its sovereignty over New Caledonia and French Polynesia, while Britain controls Pitcairn Island (with a population of 46).

4   The Fiji Labour Party and the National Federation Party formed the coalition.

5   There were in fact two coups, the first of which took place soon after the election of the Indo-Fijian coalition. On 14 May 1987 Rabuka and ten armed soldiers, faces covered with gas masks, entered parliament and at gunpoint took the people's elected representatives by army trucks to the military headquarters. On 19 May the members of the government were released after a deal was struck to avoid the likelihood of foreign intervention (by New Zealand and Australia). The second coup took place on 25 September 1987, again led by Rabuka. This one installed Rabuka as the country's leader.

6   Lapita ceramics are food vessels covered with intricate patterns. Today the term 'Lapita' refers to a Pacific Island culture beginning about 1500 BC. Lapita is a site in New Caledonia where some of the first discoveries of the ceramics were made.

7   David Stanley, *Moon Handbooks: Fiji* (6th edn), Avalon Travel: Emeryville, USA, 2001, p. 38.

8   Stanley (2001, p. 39) writes that a US threat for payment of US$45 000 'hung on Cakobau's head for many years, the 19th century equivalent of 20th century third world debt'.

9   As Winston Halapua (*Tradition,* Lotu *and Militarism in Fiji,* Fiji Institute of Applied Studies: Lautoka, 2003, p. 5) writes: 'Crucial to the colonial administration's perceived protective function was the pivotal advisory role of the establishment of the Council of Chiefs, which was institutionalised by the colonial administration.' He continues that the Council of Chiefs was not an initiative of the colonial administration, but 'a loose communal network of some chiefs' that existed prior to colonial government (p. 7). This network was subsumed by the colonial government for both strategic and financial reasons. If the chiefs were pliant, why not use their high respect in the community to control the people! And if the chiefs could govern without the costly apparatus of a British bureaucracy, all the better.

10 The ethnic Fijians are predominantly Melanesian with a Polynesian mixture.

11 Only one mine exports gold. Another mine was forced to close in 1997 due to large debts; the remaining mine is not necessarily profitable.

12 Dalo, also known as taro, is a root crop. It is under threat from the dalo beetle, and Australia and New Zealand will not allow the importation of contaminated product.

13 This is an example of the 'second best' problem in economics. We know the price of sugar is subsidised, but items of clothing (a tourist product as much as an export item) are also sold at distorted prices. Where there are numerous distortions in an economy it is virtually impossible to determine what the price of each item should be to give a 'second best' situation.

14 The inbound tourists pay Fijian nationals for the pleasure of visiting and seeing their country; hence this is equivalent to an export industry.

15 Diesel generation attracts a high subsidy.

16 The further development of hydropower is constrained by land availability and/or getting approval from traditional landowners. Notwithstanding its tropical location, droughts can be a problem in parts of the country and hence limit the potential capacity for hydroelectricity.

17 None of the suggested businesses require high levels of capital.

18 See Greg Fry, Political legitimacy and the post-colonial state in the Pacific: reflections on some common threads in the Fiji and Solomon Islands coups, *Pacific Review*, 12(3), October 2000, for an analysis of the threads common to the Fijian and Solomon Island coups.

19 Helen Hughes, A country bled dry, *Sydney Morning Herald*, 4 September 2004.

20 Hughes 2004.

6

A SMART OIL STATE LOOKS TO THE FUTURE: THE UNITED ARAB EMIRATES

Dubai has a passion for trade.

Michele Barrault

At the beginning of the twenty-first century, the world's few oil-rich states are at crossroads on the development path. Their 'black gold' is the most crucial finite resource today – but for how much longer is debatable. The conventional answer is not long – a century at the most, far less at a minimum. Only 100 years ago oil mattered little. Its refined product, kerosene, was used to produce light and the kerosene lamp was found in homes throughout the rich countries. These were still being used as recently as 40 years ago in rural Australia when I worked in shearing sheds in the country's vast rural areas.

Today it is a different world. Modern economies are literally driven by petroleum, diesel and natural gas. Yet the petrol economy is nearing its irreversible demise. While there is debate as to when this will occur, there is a strong view that by about 2010 the production of oil will have reached its feasible maximum and will start to gradually decline. Other liquid petrol sources such as shale oil, if able to be economically exploited, will considerably extend the life of the fossil fuel economy.[1]

GLOBAL WARMING AND OIL STATES

Our concern is not just with the rate of depletion of a non-renewable resource. There is the backdrop of global warming caused by climate-changing greenhouse gases that are generated by the burning of oil and other fossil fuels. Most scientists believe that the greenhouse problem is a far greater threat than previously thought. Taken together, the inevitable depletion of fossil fuels and their role in climate change will have dramatic effects on both consumers and producers of oil. So where do the oil-rich states fit into this new picture?

When the oil is gone

Consumers will have access to alternative energy sources – even though the transition to them might be rather dramatic if there is little public awareness and acceptance of the fact that fossil fuels are running out. What can the oil-producing countries do? Will their economies dry up as the oil dries up? There are enough examples of single-product economies suffering long-term decline and eventual economic destruction as either consumer tastes change or (as is the case with oil) the non-renewable resource is depleted. Nauru, with its once prosperous phosphate-based economy, is a case in point. It went from being one of the richest contries in the world on a per capita basis to being a basket case in not much more than a generation as it squandered the profits earned from a non-renewable source.

The oil rich

No two oil-rich nations are the same. They range from highly developed industrialised countries such as Norway, which for some years ranked first in the world on the human development index (HDI), until it was replaced by Iceland in 2007 and then ranked second, to developing countries at the mid to low level of the HDI such as Indonesia. In between these extremes are the small but rich countries of Brunei and the United Arab Emirates (UAE), the world's largest oil producer, Saudi Arabia, then Russia and the 'stans'.[2]

The dilemmas

Each country faces a different challenge due to its state of development, the maturity of its economy, and its political and social culture. In this chapter the focus is the UAE as it illustrates two dilemmas facing the oil-rich countries. The first is how such a country will make the transition from a natural resource exporter to a radically different type of economy. This assumes that the country's leaders understand that they have to make this change and guide this process, something requiring considerable skill. The market won't get this transition right without government oversight and involvement.[3] If a country does not want to make this transition it is doomed to economic disaster – and no nation wishes that upon itself. But some will succumb due to ignorance, shortsightedness or the greed of the rich and powerful.

The second dilemma pertains only to undeveloped countries that have become oil-rich overnight; countries that have gone from feudal or peasant economies to modern economies in a generation. The significant wealth of these countries means that at the moment they can purchase much of what they desire. If they import goods and services this is understandable and makes economic sense. But are they forever to rely on importing foreign professionals, tradespeople and workers to make up for their lack of trained personnel? The UAE is in this situation with a near total reliance on foreign skilled and unskilled workers. It can easily afford to purchase these skills. The local people need to do very little, if any, work as a consequence of the dispersal of oil wealth within their society; however, the reliance on expatriate ('expat') skills and labour is not accepted by the ruling elite as part of a sustainable long-term future for the country. As a consequence, the UAE government is attempting to substitute nationals for expats, while also seeking to convert its economy to something radically different from being simply an oil producer. This is an interesting story from which other oil-rich countries might learn.

THE UNITED ARAB EMIRATES COMES INTO BEING

The UAE is one of few oil-rich but skill-poor countries in the world. The others in this category (with the exception of Brunei) are also in the Middle East: Saudi Arabia, Oman, Qatar, Bahrain and Kuwait. All these Middle East Arab countries are members of the Gulf Cooperation Council (GCC), and are rich in oil but poor in other natural resources and trained workers, and even poorer in a scholarly class that has the freedom to put the case for a different (and better) future. Much is going to depend on the long-term thinking of the powerful sheiks in these countries.

The UAE became an independent nation on 2 December 1971. Previously it had been under British rule. Before becoming a free country it had comprised seven emirates or sheikdoms – societies governed by one ruling family. Each emirate continues to exist today but 'federal' policies are determined at a national level. The largest emirate (at 87 per cent of the total area) is Abu Dhabi, while the best known is Dubai due to its lavishing sporting events[4] and world-class shopping festival.

The UAE's current population is over four million[5] of which about 20 per cent are Emiratis. The lack of local expertise (skilled professionals and trained artisans) and the unwillingness of local recently-rich people to take on 'dirty' or even manual jobs has resulted in a vast population of both short-term and long-term immigrants. It is estimated that about one-quarter of the immigrant worker population is from other Arab countries, about half from Muslim South Asia (Pakistan and Bangladesh) and the remainder from industrialised countries such as the United Kingdom, Australia, Germany and Norway.

The largest religious grouping in the UAE is Sunni Muslim (about 80 per cent), followed by Shia Muslim (at 16 per cent) and 4 per cent Christian, Hindu and others. The UAE is a relatively strict Muslim country.

Women wear veils and their bodies are well covered. There are separate sex schools and universities for men and women, and women don't shake hands with members of the opposite sex. In this sense it is similar to the other nearby Arab countries. Go further afield into Muslim North Africa to Egypt, Tunisia, Algeria and Morocco and only a portion of the population (usually rural people) are strictly Muslim. On the other hand, nearby Saudi Arabia is much more strictly Muslim than the UAE, which sits somewhere between the extremes of Saudi Arabia and the relative freedom of much of North Africa.

The city of Abu Dhabi is the capital of the federation of emirates. It has a population of about 800 000 people, as does the Emirate of Sharjah. The Emirate of Dubai, while small in size, has a population approaching two million. All of the other emirates have small populations. The vast majority of the population is concentrated in the cities. Most of the country is desert, which today is only seen by rich Emiratis as they literally fly along the excellent highways in air-conditioned limousines.

A barren place

The UAE is one of the most barren places on the globe. It sits within the boundaries of Rub al-Khali, an area the Arabs call 'the desert of deserts'. Try to imagine that! As would be expected of such a place, there is virtually no rainfall and there are extreme temperatures in the summer months. Yet the city of Abu Dhabi, and much of the highway verge between the inland oasis in the desert city of Al Ain and Abu Dhabi, is green. This is the result of extensive irrigation using desalinated seawater and recycled grey water.

This totally unexpected landscape is just one of many indicators of the rapid and radical changes that have occurred in the past 30 to 40 years. Before exploring how these changes occurred, here is a brief description of the country today.

A RICH PLACE

In 2005 the UAE had an estimated per capita GDP (based on purchasing power parity) of US$25 515.[6] This relatively high[7] per capita income is founded on oil and gas. Crude oil production is in the order of two million barrels per day. There are large reserves, which at present levels of production (on which too much faith shouldn't be placed) could last for another 100 years. Natural gas, the fuel touted to see the world through from an oil-based economy to the hydrogen economy,[8] is being produced at 38 billion cubic metres per year.

Other than fossil fuels, the UAE produces, for both export and domestic consumption, petrochemicals, fertilisers, construction materials, boats, handicrafts and pearls. Tourism is very important. There are nearly as many tourists per year as people living in the country. The export of goods and tourism revenue are not the complete picture of the UAE's economy. Dubai is a major port in the Middle East. Fresh fruits and vegetables, clothing and footwear, pots and pans – you name it – flow into Dubai from Pakistan, India and neighbouring Iran, as well as from further afield in Asia, to be re-exported to nearby rich Arab countries. Dubai is not yet Singapore but is on the way to becoming a similarly successful port.

Oil wealth

The drastic changes you see in the UAE, and the greenery along freeways, have been accomplished by wise spending of oil revenue. Much of the country's income is under the control of the president of the UAE, who also happens to be the ruler of Abu Dhabi. On 2 November 2004, President Zayed bin Sultan al Nahayan died. Sheikh Zayed, as he was known, became the ruler of Abu Dhabi in 1966 and ruler of the federated country when it came into existence in 1971. On his death in 2004, his eldest son Khalifa bin Zayed bin Sultan al Nahayan, known as Sheikh Khalifa, became president.

An historical perspective

It is important to put the UAE and its adjoining Arabian Peninsula states into a historical and cultural perspective if we are to make sense of the UAE's present society and economy – and contemplate its future. For most of the past millennia this region of the Middle East has been 'outside the mainstream' of Middle Eastern social, political and economic change.[9] What does this mean?

It was Arab conquests that began the Islaminisation, and Arabinisation, of much of the Middle East. Some parts of the Middle East were destined to become the 'mainstream', while others were to remain backwaters. By the time of the Ottoman era, the key areas of that empire were Turkey, the Arab Fertile Crescent (Iraq and surrounds) and Egypt, but not the Arabian

Peninsula. Throughout the long stretch of history of the Ottoman Empire, the Arabian Peninsula continued to be governed by tribal chiefs who came from powerful, elite families.

As Peter Mansfield suggests, 'The Bedouin remained in their deserts beyond the reach of the Ottoman control' … 'The Arabs of today regard the four centuries of Ottoman rule as a dark period in their history'.[10] What, if anything, unified the Bedouin clan and tribal groups into either kingdoms or confederations was the Muslim religion. There was no notion of political unity as such. The lack of interest by the Ottomans in the Arabian Peninsula resulted in it having a vastly different history over the last few centuries to Turkey, Egypt and the countries of the Fertile Crescent. There was a degree of modernisation in much of the empire, except the part we are discussing – and this is a point we must not forget. Not all ex-Ottoman countries are similar. The Arabian Peninsula is the most different, the most conservative.

Ira Lapidus points out that it was only after World War II (that is, only 50-odd years ago) that the Peninsula peoples were confronted with modern technologies and ideas, while Turkey had become a modern secular state in the early part of the twentieth century.[11] This helps us understand the vastly different cultures in the area we call the Middle East. This name – the Middle East – suggests a degree of commonality and therefore is a description and would be better discarded.

OIL IS FOUND

Oil was first discovered in Abu Dhabi in 1958. However, it was not until after federation and independence from Britain that the UAE began to experience the benefits, and the shock, of vast amounts of oil income. What to do with the money was the crucial question; a question that could only be answered by traditional tribal chiefs for the simple reason that they had the power to make all the important decisions. It was their country and their oil, and their right to determine how the oil profits were spent.

Their subjects (this is the correct description) had no say. For those who consider this notion foreign, it is no different to the rule of the absolute monarchs of England before Magna Carta. And Magna Carta only slightly shifted power, giving more to the aristocracy. Democracy as we know it had to wait until the twentieth century. We tend to overlook the fact that feudalism remained a viable form of government for much of the world through to the present era. The Enlightenment and its political and economic revolutions were, by and large, contained to Europe and European settlements. A chiefly class still rules most of the Arab Muslim countries and it is the one area of the world where pre-democratic government is the norm. The most important chief in the UAE was the country's first ruler, Sheikh Zayed.

The story of the modernisation of the UAE has been told in an easy and simple manner in the personal account of Mohammed Al-Fahim in his 1995 book *From Rags to Riches: A Story of Abu Dhabi*. The title is suggestive. Al-Fahim describes how the UAE 'jumped from the 18th century into the 20th with one great leap. It went from "no tech" to "high tech" in a matter of a few years'.[12] It is difficult to conceptualise a transformation of such radical proportions. There is no written history to turn to and we have to rely on imagination and a few available photographs to understand what happened over the course of less than a generation. We are extremely fortunate to have the few personal sketches, and the fewer photographs, we have.

The Bedouin before oil

Imagine that you are a Bedouin man or woman in 1970. You would be part of a patriarchal group. The group's size would depend on whether it was a family of close kin or a tribe of related families. Go back far enough in time and you and all other members of your tribe would have a common ancestor. There could be several thousand in your tribe (or family), as there is today in the Saudi royal family. Each defined group would have as its head a sheikh, the person we would call an 'elder' or 'chief' in other tribal societies. The sheikh had to inherit the position from his father. None of this is far removed from the absolute monarchies that ruled throughout Europe and Asia for much of history; the only significant difference is that the people in the UAE today still live in this political environment.[13] Only those UAE residents who travel overseas comprehend the immense differences between their situation at home and that of modern societies. At least they can make the comparisons. People of the developed world have little to help them to visualise a feudal society.

A sucked date pit

As a citizen of Abu Dhabi, Dubai or any of the other emirates in 1970 you would be born into a specific social class and would find it difficult to marry out or otherwise leave it. You could be a member of the ruling elite or a member of a profession, such as a merchant, a fisherman, a boat builder or a farmer. If you were a member of a tribe living on the coast you would fish. In the past you or your ancestors would have been pearl divers. If you were from an inland oasis, on the other hand, you would harvest dates and transport them to market by camel – dates being the staple food in 1970. Sheikh Zayed is reported to recall a childhood when his people attempted to fool their hunger by sucking date pits. You would have drunk camel milk and, for coastal people, fish was a key source of protein.

One of the founders of the modern UAE informs us that throughout the eighteenth, nineteenth and most of the twentieth century, people lived an 'almost primitive' existence.[14] If you were a goat herder you would spend your days searching for a skerrick of edible greenery in the desert, and when your goats were fattened you would walk them to market. This miserable existence drove Sheikh Zayed to go to the lengths he did to develop his country. He experienced this 'misery first hand'.[15]

Pearls and pirates in the past

Go back a little earlier in history. In the sixteenth and seventeenth centuries, the Portuguese, the French and the English all took an interest in this part of the world, and this came to play a significant role in defining the boundaries of the new UAE when it was to eventually emerge. These three global powers contested sea routes and trade wherever there was scope for economic gain. Pirates plied their trade in the Persian Gulf waters.

An additional enticement to the area was the abundant pearls. Tribes lined up with their favorite foreign supporters, just as in the Cold War era countries lined up with one or other political bloc. In 1830 a major conflict between two of the powerful tribes – with the British being caught up in the fight – left the oyster banks destroyed. After that the pearling industry was in serious peril. For the pearl divers this could have been a relief. Gustave Flaubert, in his *Desert and Dancing Girls*, describes pearl diving: 'Two men go in each canoe, one to row and one to dive, and they go out on the open sea. When the diver returns to the surface he is bleeding from ears, nostrils and eyes'.[16] How long these poor blighters lived would be interesting to know.

A truce was reached in 1835. It was renewed every year until 1843 when a ten-year cessation to fighting was agreed to. Finally, a perpetual peace came about in 1853. Then what the British used to call the 'Pirate Coast' became the 'Trucial Coast'. The truce did not mean the situation was static. Sheikhs disagreed among themselves and, in one such disagreement, Dubai broke away from Abu Dhabi. At the beginning of the nineteenth century, the Trucial States comprised three sheikhdoms – Abu Dhabi, Ras al Kaimah and Sharjah – compared to today's seven. The families of the tribal chiefs who rule the UAE today were also the influential families in the nineteenth century; for example, in 1833 the grandfather of Sheikh Zayed became ruler of the Emirate of Abu Dhabi. This conservatism led to the continuing rule of Sheikh Zayed's family over Abu Dhabi, and hence over most of what is today the UAE.

Recalling the past

When Sheikh Zayed was ruler of Al Ain, he irrigated the area, planted trees and made it 'the greenest city in Arabia'.[17] On a grander scale, the development of Abu Dhabi, and more recently the UAE, commenced when Sheikh Zayed started to build roads, houses, hospitals and schools, and to bring electricity to all. By providing this infrastructure, he encouraged the traditional wandering Bedouins to settle down in one place. Hence, we have the few large towns we see today and the empty desert.

Mohammed Al-Fahim uses the word 'incredible' in making the point that 'less than fifty years ago our [Emiratis'] lives hung in the balance almost daily',[18] and today Emiratis want for nothing. If 40 years ago, you were sucking on date pits as a substitute for a proper meal, and today you are eating three-course meals the ingredients for which come from Australia, New Zealand and Denmark, the point is made. As long as anyone could remember the people had drunk brackish water drawn from wells and carried long distances by women. They had no electricity, no permanent houses and little knowledge of an outside world where people drove motor vehicles; flew in aeroplanes; talked to friends, family and business partners on the other side of the world; and watched television.

Al-Fahim writes of the pre-oil days:

Life in the Sheikhdom of Abu Dhabi at the time of my birth [1948] was the same as it had been in 1800 – primitive. It remained that way throughout the years of my youth.

In 1950 the population of the sheikhdom ... was still very sparse. On the island of Abu Dhabi it had reached a peak of nearly 6000 souls by the beginning of the century [1901] but by the mid-1950s had declined to about [2000] because of the collapse of the pearl industry and the subsequent emigration ... The people who lived there were of several different tribes ... The tribes intermingled more than they did in the desert ... In the desert the tribes kept to their own encampments, which were often separated from those of other tribes by several days travel by camel.

The harsh weather conditions on the coast made people reluctant to live on Abu Dhabi Island ... Women and children ... moved either to Al Ain or the Liwa while the men were out pearling ... Hardly ... five per cent of the population ... stayed in Abu Dhabi year-round because of the lack of fresh water and the poor living conditions. People only began staying throughout the year from the late 1960s onwards when businesses and offices were permanently established there ... There were few mechanical vehicles ... no roads. The northern sheikhdoms such as Dubai ... were reached primarily by boat. It still took about seven days to travel between Abu Dhabi and Al Ain [a one-hour car trip today] ... residents of Abu Dhabi island still lived in baratsi huts built of palm fronds ... a few wealthier residents and the ruling family ... lived in earth or clay houses ... We normally used mats woven from palm fronds for flooring ... sitting, sleeping ... when it was cold the family huddled in one corner of the hut to keep warm … In the summer we slept outside … where there was at least the probability … of a soft breeze … women took care of the household including cooking, cleaning and caring for the children … [and] fetching water. They carried it from the wells either in goatskin bags slung down their backs from leather headbands across their foreheads or in clay urns balanced on their heads … it was the women alone who had to bear the heavy burden of the twice-daily trips back and forth to the well … they also cared for the livestock including goats and camels … In the 1940s and 50s, when processed baby food and other modern aids for child rearing were available in the western world, there was no powdered milk or baby formula … in the whole of the Gulf … Food was scarce … mostly rice, fish, yogurt and dates … We were undernourished for the most part … Clothing was also … scarce. The men usually had a single kandoura or Dishdash … replacing it only when it was completely worn out … the women contented themselves with one or two kandouras … No one had any shoes so we all walked barefoot.[19]

Virtual illiteracy

In the 1950s there were no hospitals or schools, except for one religious school, as would be expected in a strict Muslim society. Virtually no-one could read or write, and the rate of illiteracy was 98 per cent. The sole book available was the Qur'an (Koran). Until the late 1960s herbal medicines were the only treatments to be had and, as there was no hospital until 1967, or even trained medical doctors, mortality was high. As a consequence, families were small. Given this situation, it is easy to understand Al-Fahim's 'disappointment' that the British, having been in the Emirates for almost two centuries, 'never lifted a finger to help us in the areas of education and health care'.[20] It was not unusual for colonial powers to not care much about the material wellbeing of their subjects, while they were likely to view their subjects' religious wellbeing as something to be meddled with. The British were no different in the UAE except that they were uninterested in the religious lives of Emiratis.

The fact that we know as much as we do about the UAE in its pre-modern period is remarkable for reasons Al-Fahim notes:

> *Since virtually no one could write before 1960 we have no documented historical records of our own. ... Because of this no one but those who were here at the time will ever know our perspectives on what happened ... treaties and agreements were not properly tracked and documents were more often than not, misplaced or lost ... births, deaths, marriages and the milestones of people's lives are not registered anywhere. Most local people born before 1960 can only guess at the year and date of their birth.[21]*

Every time I visit Dubai and wander around the shops, supermarkets and flash hotel lobbies, I find I have to remind myself of how recent all this is – and I come to understand once again why there have to be so many 'expats' in the country, as only they have the expertise to manage the modern infrastructure and service the visitors, most of whom have no understanding of the Bedouin past of the Emirates. Most visitors, including university academics who come to teach in the UAE, do not have an accurate picture of the recent past of this country. Its history is only now being researched – and this research will need to progress rapidly while people born from the late 1930s to the mid 1950s are still alive. They are our only source of information.

Even those who have dug deep for information know only a little about how people lived in the 1950s and 1960s, the period in which oil wealth brought about the drastic changes. Where there had been no houses, roads, schools or hospitals, by 1965, there were six schools and 528 students; by 1971, 25 schools and 7897 students; by 1993, 500 schools with approximately 300 000 students.[22]

It is worth pausing from a moment to reflect on the fact that of the Emirati population aged 40 and above today, less than 8000 have been adequately educated to play a key role in modern society.[23]

DEVELOPMENT AT LAST

Until oil was found the UAE's only export had been pearls. It is estimated that 22 000 people worked in the pearling industry; the fleet at Abu Dhabi, the largest in the region, consisted of about 400 boats in its heyday.[24] The Great Depression in 1929, and the development of cultured pearls in Japan at about the same time, led to the radical decline of the pearl-shell industry worldwide, including the UAE industry. Following the collapse of pearling, the Emirates experienced 'great poverty'.[25] By the 1950s high levels of migration from the UAE were occurring.

In 1946, Sheikh Zayed was given, by his older brother Sheikh Shakbout (ruler of the Emirate of Abu Dhabi), responsibility for Al Ain, now a very comfortable inland university city. It was here that he started to put in place his pro-development ideas. Sheikh Shakbout was a much more conservative person than his brother and would not have spent the oil money, when it started to flow, in the manner or to the extent that Sheikh Zayed did. Fortunately for Abu Dhabi and, later, for the UAE, Sheikh Zayed took over as ruler of the former and then became president of the latter.

Commenting on the first ever shipload of oil to leave the country on 5 July 1962, Al-Fahim stated, 'Abu Dhabi had at last emerged from two thousand years of obscurity to the status of an oil-providing state'.[26] The consequence was provision of land and houses – free or through very advantageous loans[27] – plus the construction of schools (for both girls and boys), hospitals, roads and the provision of electricity. These were to dramatically change Bedouin tribal society. As Michèle Barrault states, Sheikh Zayed's developments were encouraging the Bedouins to settle down gradually.[28] Agriculture was given a very significant boost through irrigation, the building of dams to harness seasonal rains and the drilling of wells. Furthermore, farmland was given to those wishing to farm, as were seeds. It was the Sheikh's prerogative to allocate this land and give away money. In an absolute monarchy, it was in some sense all his to give away. All in all, these allocations were devices to convince Bedouins to settle in one place, whether it be Abu Dhabi, Al Ain or any of the other towns or cities.[29]

Desalination and the first hotel

In 1961, two small desalination plants were commissioned which were to prove vital for an oil-rich, water-poor country. Abu Dhabi's first modern hotel, the Abu Dhabi Beach Hotel, was constructed in 1962 – a 'giant' at 25 rooms and with electricity! Today the upmarket Abu Dhabi Sheraton hotel is located near the site of the original hotel. In 1963, the first post office was opened and a telephone system was installed. Abu Dhabi's first hospital, the Central Hospital, was opened in 1967. Obviously, there were no local people trained as medical doctors, nurses, managers or even cleaners. Given the extent of illiteracy in the society then, parents were not instilling in their children a desire to do well at school so as to go to university to become a medical doctor or the like. Even today, virtually all hospital staff (from professionals to cleaners) are expats. Emiratis have yet to gain either technical skills or a work ethic suited to the routines of modern hospitals.

It was only a little over a generation[30] ago that the UAE got its first university and that was in Al Ain in 1977. By the early 1990s, the number of hospitals had increased from one to 42, of which 33 are what we would call public hospitals (that is, they do not charge patients). Before the opening of the UAE's first university, a very small number of Emiratis obtained a university education by travelling abroad, but it is difficult to find evidence of the numbers or whether any females did so. Given the quite recent shift from a male-dominated tribal society, it is worth pointing out that today more females attend university in the UAE than males and twice as many women graduate.[31]

Women in development

Don't be surprised to find women driving their own cars and doing their own banking, some even wear European clothing. Geraldine Brooks tells us to expect to see 'women soldiers, their hair tied back in Islamic veils, jumping from helicopters and with shoulder assault rifles'.[32] Across the border in Saudi Arabia, women are not even allowed to drive a car. For a traditional Muslim society recently emerging from tribalism, the UAE is different!

Emiratis live in 'splendid houses that look like palaces',[33] Barrault informed us in 1993 and today, over a decade later, the palaces are larger and there are many more of them. The following statistics from the 2005 HDI paint a picture of a prosperous, modern society (at least in terms of material goods and services). Average life expectancy is 78.3 years, the adult literacy rate is 88.7 per cent, the public expenditure on health is 2 per cent of GDP[34] and health expenditure per capita is US$503.[35] There are 1000 mobile telephones per 1000 people.[36]

Most of these remarkable achievements have been attributed to Sheikh Zayed: 'Here was a man who had never been to school, a man who lived a nomadic Bedouin life since his youth. Yet he both conceived and implemented the ideas that would be the foundation of the entire infrastructure of modern day Abu Dhabi [and the UAE].'[37] Undoubtedly expatriate expertise was essential in the realisation of these results. As Al-Fahim writes, 'the local population was simply not qualified yet to complete the tasks that needed to be done'.[38] Today there is recognition that reliance on expats is not sustainable. Notwithstanding the government policy to bring people in from other Muslim countries (as well as Arab people), to presumably minimise religious and cultural differences with the UAE people, there have been problems.[39] It was not – and is not – all rosy for the expats. They are not allowed to purchase land, establish a business or buy shares without a local partner.

Emiratisation

'Emiratisation' is the official policy in the UAE today – local people are to be trained to take over the skilled jobs of the expats but progress on this front is awfully slow. There is little in the way of research into the attitudes of the Bedouins about matters that consume modern societies, such as organisational structures and processes, timekeeping and punctuality, and general work behaviour – or at least that I am aware of. While there are reams of books on Muslim culture and religion, this is a different issue. They don't inform us of the difficulties that Emiritisation may cause.

Sandra Mackey has written extensively on Middle East matters, and while her observations are not specifically about the UAE, her general observations about Bedouin culture and attitudes, particularly in Saudi Arabia, might apply in thinking about the process of Emiritisation. She describes the Bedouins' 'deep love of freedom'[40] and fierce independence. As a consequence of this, she believes that they 'do not plan and organize',[41] and that time is not important to them. She suggests that they have an aversion to labour that is 'buried deep in Bedouin tradition'[42] – a point I find interesting but not very convincing, given the hard work that most Bedouins must have faced just to survive in such a demanding desert environment.

How the rich live

If Mackey's assessment is correct, it explains the heavy reliance on expat labour for manual work or occupations held in low esteem. But perhaps this reflects nothing more than the attitudes to work of the rich in all cultures. The Emiratis live how the rich have lived for millennia in all societies – with a servant class doing the manual and dirty jobs. That the oil wealth has been shared among the population of UAE nationals perhaps has resulted in an expectation that wealth can be obtained without work. However, the fact that there is a large, enterprising business class in the UAE, only a generation after the change from a tribal society, suggests otherwise, as does the fact that universities have high enrolments, particularly for women. The new Bedouins are entrepreneurs and they are not going to simply live on their capital. This, of course, is a good thing because sustainable development demands that future income flows have to be factored in to today's decisions.

POSITIONING FOR THE POST-OIL ERA

No other Middle Eastern or Asian oil state has moved so decisively to position its economy for the post-oil era than the UAE. The UAE has, at the very least, two things going for it: strong leadership and geography.

I tend to compare the UAE with Brunei when I think of the future of the small oil states. They have much in common: both are very rich countries with small populations; both are absolute monarchies with only notional parliamentary representation. One of the frustrating consequences of this is that it is impossible to analyse their economies in any meaningful way. For that you need publicly available data and transparent decision-making. Brunei, as does the UAE, relies heavily on foreign workers, both skilled and unskilled. In the UAE, more so than Brunei, this leads to uneven service. There are far more overseas workers in the UAE and many are more poorly trained than in Brunei. While I hesitate to make claims based on personal experience, the majority of professional drivers I have been assigned in the UAE have had considerable difficulty finding simple locations. Some could not read English – this in a country flush with business types all speaking the language of commerce, English.

Both the UAE and Brunei are strictly Muslim, although the long-established and prominent Chinese population in Brunei is not obviously

Muslim. The population mix in Brunei has a moderating effect on the Muslim culture. The most striking similarity of the two countries is that the rulers of both are spending part of their enormous wealth to keep the people happy. High-standard housing, good infrastructure, a fun park in Brunei, universities and much more benefit the locals. Both countries have, or are building, palace-like hotels. (In Brunei's case, one presumes the guests will be the few super-rich who visit the country to meet the sultan.)

Hartwick rule

Not for the Emiratis a narrow interpretation of the Hartwick rule,[43] which, if followed strictly, would see them make serious efforts to invest in future energy sources, replacing a depleting energy source (oil and natural gas) with, say, solar power. Rather, their strategy is based largely on the favourable location of the UAE. Like Singapore, geography matters a lot, for what is modern Singapore but a large shop and a very large port with a world-class airline? What is the UAE, if not this?

There already is a Singapore in Asia. There is soon to be one in the Middle East. Yet is an Arab 'Disneyworld' – with underwater restaurants, indoor ski slopes, artificial islands by the score and, next, the world's tallest building (because this is where Dubai and Abu Dhabi in the UAE are heading) – a sustainable bet?

1   With the high prices for conventional fuels, a range of cheaper alternatives will be able to compete. Where in the queue of these alternatives shale oil sits at the moment is not easy to answer.

2   The 'stans' include Kazakhstan, Kyrgyzstan, Uzbekistan, Tajikistan and Turkmenistan.

3   This is not to say that, without government guidance, wise investments would not be made from the profits of oil. However, overall unwise investments driven by short-term profit-seeking and, more than likely, under-investment will probably be made.

4   Some of these events are held in the neighbouring emirate of Sharjah, where alcohol is not legally available. I suspect that some tourists spend as little time as possible in Sharjah!

5   This figure is a projection for 2008, based on 2005 census data, but includes a high estimate of non-citizens.

6   UNDP, Human Development Report 2007/08 at http://hdr.undp.org/en/statistics/.

7   Norway, which in 2005 ranked second in the world on the UNDP's human development index (and is another oil-rich country) had a per capita GDP (based on PPP) of US $41 420. Australia, ranked third in the world in 2005, had a comparable income of US$31 794 (UNDP, Human Development Report 2007/08 at

http://hdr.undp.org/en/statistics/).

8   Hydrogen is one of a number of possibilities to replace fossil fuels, and is placed to be our major fuel in about 40 years (Chapter 3 explores this in greater detail).

9   Ira M Lapidus, *A History of Islamic Societies* (2nd edn), Cambridge University Press: Cambridge, 2002, p. 566.

10   Peter Mansfield, *The Arabs*, Penguin. Harmondsworth, UK, 1992, p. 65.

11   Lapidus 2002, p. 567.

12   Mohammed Al-Fahim, *From Rags to Riches: A Story of Abu Dhabi*, London Centre for Arab Studies: London, 1995, p. 137.

13   The UAE is not unique in this regard. The other Gulf Corporation Council nations remain tribal. There are also numerous countries (in Africa, Asia and Oceania) where the status, authority and power of chiefs, elders or warlords are considerable.

14   Al-Fahim 1995, p. 21.

15   Michèle Barrault, *Regards: The United Arab Emirates* (trans. T Lehane & A Pargoud), Editions Michel Hetier: London, 1993, p. 19.

16   Gustave Flaubert, *Desert and Dancing Girls*, Pocket Penguin 17: London, 2005, p. 54.

17   Barrault 1993, p. 19.

18   Al-Fahim 1995, p. 15.

19   Al-Fahim 1995, pp. 53–7.

20   Al-Fahim 1995, p. 58.

21   Al-Fahim 1995, pp. 58–9.

22   Barrault 1993.

23   Children from better-off families were sent overseas to study, hence the number of educated middle-aged people is probably higher than the data suggests.

24   Barrault 1993.

25   Barrault 1993, p. 18.

26   Al-Fahim 1995, p. 93.

27   Enough money was given to people to destroy their baratsi huts and to build a modern house plus establish a business. In addition, each citizen got three or four plots of land for housing and business purposes (Al-Fahim 1995, p. 140).

28   Barrault 1993, p. 19.

29   Barrault 1993, p. 30.

30   A generation is taken to be 25 years.

31   Barrault 1993, p. 67.

32   Geraldine Brooks, *Nine Parts of Desire: The Hidden World of Islamic Women*, Anchor: Sydney, 1995, p. 5.

33   Barrault 1993, p. 30.

34   Compared to 8.1 per cent in Norway, 6.5 per cent in Australia, 2.2 per cent in Egypt and 1.0 per cent in Indonesia.

35   Compared with US$4080 in Norway, US$3123 in Australia and US$258 in Egypt.

36   The cellular subscribers level in the UAE is more than Australia at 906 but less than Norway at 1028.

37   Al-Fahim 1995, p. 135.

38   Al-Fahim 1995, p. 135.

39   See Al-Fahim 1995, p. 172.

40   Sandra Mackey, *The Saudis: Inside the Desert Kingdom*, WW Norton: New York, 2002, p. 182.

41   Mackey 2002, p. 236.

42   Mackey 2002, p. 183.

43   The Hartwick rule states that (at least) part of the profits (or rents) earned from the exploration of a non-renewable resource such as fossil fuels is to be invested in replacement earning assets, such as renewable energy sources.

7

RUSSIA AND OTHER COUNTRIES IN TRANSITION

The end of the Cold War …
sudden and totally unexpected …
changed the face of civilization and the fates of nations.

Daisaku Ikeda, in conversation with Mikhail Gorbachev

In this chapter we take the opportunity to explore whether or not we can learn anything about the human condition and development by scrutinising countries that have been 'shocked' into radical change, such as Russia, Bulgaria, Czech Republic, Poland, Slovak Republic, Ukraine, China and Vietnam. There are lessons about human psychology and about values and attitudes to political philosophies and economic realities to emerge from a retrospective on the period that followed the dismantling of the Soviet Union in 1991. We all face the dismantling of the carbon economy.

No slow evolutionary process was allowed to occur in the transition of Russia and some of its former satellite countries (even though this was an option), and hence the impacts on these societies are rapid, raw and reliable signals of societies undergoing immense change.

What values and attitudes did the people of these countries bring to the table, so to speak, when they set out to establish a new form of governance? Were their values and attitudes significantly influenced by their time under communist government? Did 70-odd years of communism change people?

Here we concentrate on what has happened to these peoples in their transitional phase since the 1990s. The only realistic approach is to delve into whatever information we have of the trials and tribulations of these peoples in transition, with only a cursory glance at their history where this might assist our understanding. We are constrained by the lack of relevant and reliable data, even in considering recent history, so this chapter relies on recent survey data from the Pew Research Centre[1] and the data on the rate of progress (or regress) of the countries in transition available in the human development index (HDI) reports.

Some of the ex-communist countries have already completed the journey from Soviet society to twenty-first century capitalism and are now members of the European Union. Others have either made slow progress or have regressed. There is a range of reasons for the differing pace of change – and some of the differences are surprising.

BEFORE AND AFTER

We commence our analysis with the recent history of the spread of development of our chosen countries. Table 1 shows their HDI rank as of 2005. While this year has passed, it is an appropriate reference point being close to a decade and a half after the collapse of the Soviet Union – time for significant changes to occur if they were going to. Let us consider what happened over this period.

In 2005, the Czech Republic, ranked 32 out of 177 countries, was the highest ranked of our chosen countries on the HDI. Either side of it was oil-rich Brunei Darussalam at 30, Kuwait at 33, and at 31, Barbados. This puts the Czech Republic in the UNDP's 'high human development' category. For those looking for another benchmark of development, the HDI ranking of each country is closely correlated with the gross domestic product (GDP) per capita ranking for the same year. The Czech Republic sat at 34 in the world in terms of income earned per person.

Table 1 also shows changes in per capita income from periods before and after the collapse of communism in selected countries. There are startling increases in income in our chosen countries from 1990 to 2003; however, some countries such as Vietnam had an extremely low base income in 1990. These numbers should not be taken to show that the transition from communism was the main, or only, reason for growth; however, it has been important – even in Vietnam and China, where communism remains more in name than in practice.

Table 1: Selected countries in transition: trends in income[2]

COUNTRY	HDI RANK 2005	GDP PER CAPITA (US$ 1995)				GDP PER CAPITA 2005 (US$)
		1975	1980	1985	1990	
Bulgaria	53	N/A	1544	1805	1994	9032
China	81	N/A	N/A	N/A	1341	6757
Czech Republic	32	N/A	N/A	5679	6124	20 538
Poland	37	N/A	3407	3276	3370	13 847
Russia	67	2969	4246	4024	4262	10 845
Slovak Republic	42	N/A	N/A	4218	4445	15 871
Ukraine	76	N/A	N/A	N/A	1555	6848
Uzbekistan	113	N/A	N/A	N/A	1555	2063
Vietnam	105	N/A	N/A	213	239	3071

Poland, the Czech Republic and the Slovak Republic were only marginally affected by the collapse of the Soviet Union, having gone through their own revolutions in an earlier period. They had achieved a reasonable standard of income by the mid 1980s. Nevertheless, in the period since the collapse of the Soviet Union, all the selected countries achieved solid economic growth. The weakest performers were Uzbekistan and Russia itself.

For poor countries, changes in per capita income are reliable indicators of a people's state of wellbeing. However, there are other indicators that better illustrate progress, or as the case may be, regress. One such measure is life expectancy at birth. What happened in a generation to the people living in our selected countries is shown in Table 2. The progress made in Vietnam and to a lesser extent in China stands out. Recall that a generation ago Vietnam was just emerging from decades of war[3] and hence the low life expectancy in the period 1970 to 1975 is easily explained. We should note that China's economic growth success story is reflected in increased longevity.

Table 2: Selected countries in transition: life expectancy at birth[4]

COUNTRY	LIFE EXPECTANCY AT BIRTH (YEARS) 1970–75	LIFE EXPECTANCY AT BIRTH (YEARS) 2005	INCREASE/ DECREASE (YEARS)
Bulgaria	71.0	72.7	+1.7
China	63.2	72.5	+9.3
Czech Republic	70.1	75.9	+5.8
Poland	70.5	75.2	+4.7
Russia	69.7	65.0	-4.7
Slovak Republic	70.0	74.2	+4.2
Ukraine	70.1	67.7	-2.4
Uzbekistan	64.2	66.8	+2.6
Vietnam	50.3	73.7	+23.4

THE RUSSIAN STORY

Russia is a giant country. It is both a European country (west of the Ural Mountains) and an Asian country (to the mountains' east). We can trace its history, its earliest known people, back to 20 000 BC; however, our task is not to attempt to describe its history or its vast geography. Many others have done that. Our focus is the failed challenge – to date – to build a sustainable society after the collapse of communism. What follows is history in television grabs – because I want to get to the main story quickly.

The Rus form the Russian nation

The one brief historical point I should make is that the first Russian nation ('state' is probably the more appropriate word) developed because of trade on its major river routes. Trade made Russia and has the potential to remake it into a modern superpower. The first Russian state was formed when the Vikings (mainly Swedish, although some were from Denmark and Norway), known as Varangians by the Slavic people, came across the Baltic Sea into the rivers of Russia (such as the Volga) and eventually reached Byzantium (modern Istanbul). Thomas Noonan describes these people as 'explorers, adventurers, rulers, merchants, mercenaries, farmers, political exiles, and raiders'.[5] Some settled and the Rus state was formed. The conventional view is that the word 'Rus' is derived from the West Finnic term for Sweden, 'Ruotsi'.[6] It should be noted that today's Russia does not include all the areas that were subject to Viking influence or rule; for example, the Viking city of Kiev is in the Ukraine. The major Viking cities were Novgorod, Smolensk and Kiev. Together they formed a loose grouping of city-states.

The rise and fall of the Soviet Union

In 1917 the Soviet Union came into being. Russia, plus a grouping of 'republics', formed an extremely tight union based on a form of state socialism, commonly called 'communism', but it was far from what the promoters of communism – or for that matter, socialism – believed either should be. The Soviet empire started to crumble in the late 1980s. In 1985 Mikhail Gorbachev became the general secretary of the Soviet Communist Party and President of the Soviet Union. He was to be the last president. In 1990 he was awarded the Nobel Peace Prize for his efforts leading to the end of the Cold War.

Glasnost and Gorbachev

In recent years Gorbachev has reflected on what happened after he embarked on the process of *perestroika* and *glasnost*; that is, openness and a degree of freedom and liberalism, both politically and economically. It was a fascinating time, as Gorbachev reflected in a discussion in 2005 with Daisaku Ikeda, the well-known Japanese Buddhist philosopher and educator. Ikeda, in these few words, neatly summarised Gorbachev's influence on the world.

> *Your appearance in the political arena in the 1980s was truly fateful for world history. Perestroika, of which you were the father, led to the end of the Cold War, the democratization of the totalitarian communist regime in Russia. These events were sudden and totally unexpected. They changed the face of civilisation and the fates of nations, ethnic groups, and individual human beings.*[7]

We witnessed throughout most of the twentieth century in the Cold War a battle between two competing models. The competitors went by different names, but let us call them 'capitalism' and 'communism', although there has never been anything approaching a communist society as defined by Karl Marx. The Soviet model was a form of state socialism, according to numerous political commentators; however, this term does a disservice to socialism. And capitalism comes in numerous guises, very few being the 'robber baron' style of the early USA and modern Russia! The contest between these two economic and political systems was one of the most critical in human history. No sooner had this contest, which had taken place over more than 70 years, drawn to a close than a new period got underway. How would the countries previously under Soviet rule (including Russia itself) adjust to the demands of modern life in a global market economy? How would they deal with the enormous backlog of environmental problems that their rapid industrialisation under communism had caused? Could they rebuild societies based on the new principles of sustainable development, fairness, concern for the future and care for the environment?

Russia goes backwards

Our major interest in the data is what it shows when a long-established system of government collapses and is not immediately replaced with a well-functioning alternative. Where this was the case, whatever social capital (well-established and much-used government institutions and, at the individual level, trust, honesty and concern for others) existed under the old system was lost. Data in Table 2 shows the standout country is Russia. A decline in life expectancy over such a short period is unheard of in countries as advanced in technology, medicine and overall wellbeing as Russia was in the late 1980s. Had something similar happened in the USA, Germany or Australia there would have been great social unrest and governments would have been defeated at the polls. Declines in life expectancy are normally associated with major civil wars and failed states, not in military and economic powerhouses such as Russia.

What was the reason for this decline? We know that Russia suffered an enormous breakdown in social capital when the Soviet Union finally collapsed in December 1991. Why? Moscow adopted 'the neoliberal economic model as fast as possible'[8] and allowed local clans, or the mafia, to take over the most lucrative sectors of the economy such as oil, tobacco, caviar and arms. The conventional wisdom is that many of the new robber baron entrepreneurs were ex-Communist Party heavyweights, as they were in the best position to rapidly fill the gap left by the collapse of the state enterprises. The tragedy of so-called communism is that it had not led to the formation of a class of honest public servants.

The nation's resources were privatised – sold for a song if not given away – and to further exacerbate the decline in overall wellbeing there were literally no institutions – financial, legal or industrial – that were able to modify the practices of the new robber barons. It was as if the banks, the tax collectors, the police – in fact, all organisations with any authority in the previous Soviet period – were relegated to the dustbin without thought of what should take their place. Criminals did!

Most of the European countries in Tables 1 and 2 showed a marked improvement in per capita income and life expectancy and none of them experienced the social and economic trauma that Russia did. The degree of social and economic disturbance in Russia correlates with a breakdown in health care, nutrition, increased alcoholism and crime, all of which shorten lifespans.

A superpower had lost its soul and the understanding its citizens had of its reason for existence. Rather than become angry and determined to quickly fashion a new future, many gave up hope. It is hard for us to imagine just how radical and quick was the dismantling of Soviet society and its cumbersome communist, but nevertheless functioning, economy.

Vladimir Putin, Russia's second post-communist president (and current prime minister), has described the manner of the change as the 'greatest geopolitical catastrophe of the last century'.[9] He is right. Many of the devastating shocks that happened to Russia occurred during the administration of Boris Yeltsin that immediately followed the Gorbachev government. Key strategic assets – natural resources such as oil, gas and minerals, on which the economy was based – were sold at ridiculously cheap prices to private firms. That this could happen leads me to ask just how committed were the communist bureaucrats to socialism (of any kind)? This is how Richard Pipes describes their actions and attitudes.

> *It is significant that when the Soviet government disintegrated in 1991, the supposed guardians of ideological purity – the* nomenklatura *– gave up without a fight and pounced on the country's natural resources and manufactures, stripping them bare, under the guise of 'privatization', for their personal benefit. This would scarcely have happened if the apparatus had indeed been committed to Marxist–Leninist ideology.*[10]

The recent Soviet experience has destroyed much and the Russian bureaucrats who became robber barons overnight should have to explain themselves. They were not true to the class they were supposed to support.

A loss of faith and face

The immediate aftermath of the Soviet collapse illustrates just how far from a socialist ideal Russia had come. Even though most Western socialists had come to realise that the Soviet experiment had gone horribly wrong with Stalin, and that the efforts by Khrushchev and others to reform the system had not gone far enough, some held out hope up until 1990, particularly with Gorbachev, that eventually some good would come out of the system. Yeltsin destroyed any prospect of a reformed democratic socialist Russia.

One character who grabbed much of the Russian people's wealth was Mikhail Khodorkovsky. He became the country's richest man, owning massive oil fields, but today he is in jail for fraud and tax evasion and oil is being re-nationalised. Putin has, it seems, a clear eye to both the economic and strategic value of oil and gas and used these resources to commence to restore Russia's geopolitical power. Putin asserted that these reserves are a force equivalent to the once-powerful Red Army and, in an oil-hungry world, he might just be onto something.

The wrong road

It could have been otherwise in Russia. At the time of the collapse of the Soviet Union there were two competing views of how to facilitate Russia's transition from state socialism to something different. The model that was talked about most (in Europe at least) was the Scandinavian model of a liberal social democracy in which there was an important part to be played by the state in taxing and spending, protecting the environment and investing in natural resource projects for the future. The idea was to gradually dismantle the old (communist) state bureaucracy (which controlled production of both producer and consumer goods) and replace it by markets and the ancillary institutions such as consumer protection laws, commercial law and sound, responsible banks. All these are prerequisites for an efficient market. Note that the state was to be *gradually* replaced.

Secondly, under this proposal the existing social welfare system in Russia would be remodelled – not disbanded – so as to reflect Scandinavian practice. There was no need to sell off (or blindly give away) state enterprises such as those involved in oil production. Like Norway's Statoil, they would remain in government hands and generate the money needed to provide for a sustainable economy. It was all very simple. The model had been applied and worked well in Scandinavia for generations. However, some people got in the way of a sensible solution for a Russia at the crossroads. Right-wing ideologues were intent on delivering Russia and its empire to market speculators. They had ready cheerleaders in the defunct Soviet system.

The short, sharp shock

The alternative model to that of a guided and gradual move to social democracy was the 'short, sharp shock' – destroy the old state structures overnight and allow a new form of economy to emerge from the rubble. The proponents of this policy expected that the result would be a laissez-faire capitalist economy – ripe to be plundered by Western resource interests. This was the model favoured by the economic fundamentalists. These are the people Joseph Stiglitz, an American economist and one of the more astute judges of what went wrong, calls the 'Washington consensus', basically the right-wing economists and politicians.

The drop in life expectancy in Russia was the most telling statistical result of the fundamentalist project. Poverty, crime, unemployment and a major loss of community spirit resulted, each producing their own sad statistics. The end result was to cheat the average Russian of nearly five years of life. Let the economic fundamentalists measure that in their calculations of economic welfare! The shock, which was to be short in duration, has not been, and continues today, although there are some signs of a turnaround in the first decade of the twenty-first century.

The failed transition to a market economy

Just as we have learned – or should have learned – from the failed experiment of state socialism in the Soviet Union and China, so can we learn from the failed Russian transition to a market economy. This is not to say the Russians won't eventually recover and become a stable, relatively rich and happy country. The people of Russia, as much as any other people, surely deserve a sustainable, healthy and wealthy future.

During the Soviet period, economic growth based on manufacturing, collectivised farms, forced savings and very good technical education took the country only so far on the road to development. We can't use the adjective 'sustainable' in terms of the Soviet era, as much environmental harm was done. The Soviet managerial class were environmental vandals and here is an illustration of how. Russia's total area is 17 075 200 square kilometres, of which half is temperate forest; however, less than 2 per cent of this was, and is, protected. This has contributed to a number of animals being under serious threat of extinction: four of the big cats, seven species of ungulates, seven species of bats, six species of reptiles and 38 bird species. These few statistics suggests that much has to be done. Most of the damage was done in the Soviet era, but the shock treatment handed out more recently has not helped the environment.

Russia as a special case

Maybe Russia is a special case. We can be more optimistic about some of the other countries in transition. The values and attitudes of their peoples tend to differ from those of the Russians. To these people we turn next to attempt to understand their social characteristics and the role they play in societal change and, in particular, progression to an environmentally and economically sustainable society.

The human disposition to change, or not change, will determine if we can achieve sustainable development. An easy transition from a rigid state-socialist system to a market economy would suggest that people are remarkably adaptable and resilient enough to cope with drastic lifestyle changes. The middle class of the world could be – and probably is – looking to a drastic lifestyle change due to the build up of climate-changing greenhouse gases. How adaptable will we be? Just as we look for resilience in natural environments so do we in societies. Human resilience will be increasingly necessary as we approach the end of the oil age, and the onset of climate change and other environmental shocks.

In contrast to resilience, a tortured transition, like the ongoing Russian case, suggests serious difficulties when a radical change is required of humans. The Russian experience certainly does not promote optimism as it showed that humans are not capable of quick and radical change. We are slow in making societal changes – and are rapid, even careless, in adapting to technological changes. This implies that we should start today to make the adaptations in lifestyles that will be necessary to cope with climate change and the depletion of fossil fuels.

The environmental legacy

The fact that Russia has been sent backwards means that there will be many years, probably decades, before it will have the financial means to clean up the pollution and general environmental degradation that has occurred over a period of 70 years of industrialisation at any cost. Few would have forgotten that day in April 1986 when the nuclear power plant at Chernobyl released its deadly poison. Farmers as far away as Cumbria in Italy were forced to cease grazing sheep, and acres of land near Chernobyl still remain off-limits. On the positive side, Russia has got oil (and other valuable natural resources), and if these resources are returned to the government sector, significant income will be generated; some of which must be used for environmental restoration. It is not just the pollution that requires urgent attention, but also the decline in health standards and the general decline in social capital.

In 1991, the Russian leadership faced two options. It chose the wrong one. The Scandinavian model is still waiting to be tried; however, the egg has been broken and scrambled and, hence, the task won't be easy. To meet intergenerational equity criteria, Russia needs to follow in the footsteps of Norway in how it uses its oil wealth – part of the profits must be invested in future income-earning businesses and the oil income must not be squandered on today's consumption or frittered away by the new super-rich oil mafia.

THE FREE MARKET

For each of our chosen countries the change in their economic situation has been fashioned by the fundamental transition they have undertaken or are undertaking. To varying degrees all have become 'free market' economies, even the communist-in-name states of China and Vietnam. The Pew survey asked respondents in each country whether or not they were better off in a free market economy, allowing the caveat 'even though some people are rich and some are poor'. Vietnam topped the list with 98 per cent responding that they were better off. This is a startling finding in a country still communist in name, with a people who fought valiantly to stop the capitalist, Catholic-dominated south from controlling the whole country. Today, virtually without exception, the Vietnamese state that they are better off living in a market economy. Make sense of that!

Vietnam was followed by China with 90 per cent of respondents saying they were better off. However, if we look at the gradation from 'a lot more' (better off), 'somewhat more' (better off) to 'only a little more' (better off), only 19 per cent of the Chinese respondents said 'a lot more', compared to 61 per cent of Vietnamese. Yet, this needs to be treated with considerable caution as the samples were disproportionately drawn from urban areas where the new middle class is wheeling and dealing in the manner of the best – or worst – capitalists. In both Vietnam and China there has been a rapid development of a middle class and this itself has brought changed attitudes and a significant degree of optimism by the winners in the growth stakes. What the much poorer rural peasants in these countries think and experience is likely to starkly contrast to the survey findings. Yet, we simply don't know for certain how different their views are – the data does not exist.

Following these two Asian countries in the rankings are the Ukraine (with 88 per cent better off), the Czech Republic (87 per cent), Uzbekistan (86 per cent), the Slovak Republic (83 per cent), Russia (78 per cent), Poland (76 per cent) and Bulgaria (63 per cent). Of this list, Bulgaria is furthest removed from being a modern market-based economy and this is the likely explanation for its people to not rate a market economy highly – they hardly know what a market economy is.

The pace of change

Free market economics, the availability of foreign movies and television, as well as the other significant institutional changes, are what cause people to either like or dislike the 'pace of modern life' in their new societies. Consider these survey results: the Vietnamese ranked first in liking the pace of change, sharing top place with the Uzbeks (79 per cent); next came China (65 per cent), the Slovak Republic (57 per cent), the Czech Republic (53 per cent), Bulgaria (50 per cent), Ukraine (42 per cent) and equal last were Poland and Russia (35 per cent). The people of the Ukraine, Poland and Russia stand out as not liking the pace of change, while those from Vietnam and Uzbekistan clearly appreciate the rather rapid pace. Some of these findings are truly puzzling.

The Russian respondents' opinions are understandable, given what we have already noted about the very real troubles (to put it mildly) the country went through and is still experiencing to a lesser extent in its transitional phase. Poverty (unheard of under the communists, at least from the 1960s), high crime rates and decreasing longevity must be strong reasons for the Russians not liking the pace of change. The Ukrainian people (who were close to the Russians in the old Soviet Union) had similar experiences to the Russians and this may explain their attitudes. It is not clear what to make of the Polish response as its people experienced a much longer and smoother transition than Russia. However, as we will see, the answer to another question provides some possible pointers. Keep in mind that Poland is different in one vital respect to other ex-Soviet European countries – religion.

LOSING LIFESTYLES

Respondents to the Pew survey were asked whether or not their traditional way of life was being lost. Polish citizens topped the list with 79 per cent responding that their traditional lifestyle was being lost, followed by 75 per cent of Slovak citizens. Let us explain this immediately. Poland has remained a traditional and conservative Catholic country. This is the vital difference referred to previously. Even throughout Poland's socialist period, its citizens retained their religion. Timothy Byrnes points out that the Catholic church was 'the symbolic substitute for … civil society during the darker decades of communism'.[11] This certainly is part of the reason; however, there was another strong influence on Polish values and attitudes. Poland is the least cosmopolitan of the countries we are considering, as the Poles have tended to keep to themselves. It is not necessarily the case that a cosmopolitan society is going to be less conservative than one that is ethically and culturally narrow, but it is the general rule, and Poland fits the mould.

Conservative Catholicism is likely to be the determining factor in Poland. Poles don't appreciate the challenges to their long-held views that come from foreign television and tourists and the variety of other intrusions from more open societies. The libertarian West is not for the Poles. The degree of Catholic conservatism in Poland is evident in the fact that a treaty between the Holy See and Poland (the Concordat) was agreed to in 1993. It is a moot point as to whether or not the treaty imposes the Catholic Church's will over the Polish people and their state[12] – dominance by the church was intended. In Poland, church and state combined make for an ultra-conservative European society and one would have to go back to France before 1789 to find church and state so powerfully combined.

The Slovak response to the question on lifestyles again reflects the social and economic conservatism of its people. The Slovak Republic is also a Catholic country, while its former partner, the Czech Republic, has a very high atheist population. The Czech part of the old nation, Czechoslovakia, was less conservative before the Velvet Revolution of 1989 that overthrew the communist government, and hence any differences in lifestyles in the transition period would not be as noticeable there as in the Slovak Republic. The fact that Catholicism remained such a strong social force in the Slovak Republic throughout two generations of Marxist education is a puzzle yet to be understood.

The order of the other countries in concern about loss of traditional lifestyles was as follows: Russia (72 per cent), China and Bulgaria (68 per cent), the Czech Republic and Ukraine (66 per cent), Vietnam (57 per cent) and, last, Uzbekistan (38 per cent). The relatively low score for Vietnam suggests that, at least for city-dwelling Vietnamese, the influence of visitors and foreign television is being accommodated and probably welcomed.

Uzbekistan as a special case

Uzbekistan is a real surprise. Does geography help explain its peoples' attitudes? It is not in Europe and it is not in the dynamic, economically vibrant part of Asia; rather it sits north of Afghanistan and south of Kazakhstan. This, we would think, is probably a conservative and troubled part of the world. The influence of modern, foreign ideas in this part of the world is likely to be minimal.[13] The only major outside influence in this area in the past century has been from Russia, and then the Soviet Union, when it came into existence. More recently, since the collapse of the Soviet Union, the USA has been able to develop relationships, both military and commercial, in Uzbekistan. Do the local people believe these relationships to be highly desirable? Is this what explains the survey results? Yet, the vast majority of Uzbeks are Muslim, which might suggest conservative social attitudes. In early 2005 there were serious revolts by armed Muslim gangs in Uzbekistan and the future stability of the country is in question. The survey results remain unexplained (maybe they are not representative) and the country remains an enigma.

CONSUMERISM TO THE FORE

If we turn next to the influence of consumerism and commercialism on how a people in transitional societies perceive change, we find that a relatively high proportion of the Uzbekistan population (57 per cent) believes that modern influences are not a threat. Again, a result for which there is no obvious explanation, unless the influences of mass consumption and commercialisation are yet to have an impact. This is probably the case.

Only one other nation, Vietnam, had a higher percentage of people (66 per cent) who believed that there was no threat from consumerism and commercialism. It could be argued that if consumerism and commercialism are nothing more than 'doing business' (whether buying from a stall or repairing a motorbike in a garage), the urban Vietnamese are very much part-and-parcel of the business world, and have been for a long time. In which case, these are not novel concepts and are therefore not a threat. Whether the Vietnamese would feel the same way about commercialisation on a grand scale (think Las Vegas) is another matter.

After Vietnam, the order of countries whose populations believed modern consumer forces are not a threat is: the Ukraine (51 per cent); Bulgaria (46 per cent); the Czech Republic, Russia and the Slovak Republic (all 45 per cent); and Poland last at 32 per cent. Taking into account non-responses, only the Czech and Slovak Republics and Poland had more people believing consumerism and commercialism were a threat to their culture than not.

The response for the Slovak Republic and Poland are consistent with other conservative responses in the survey results, while that of the Czech Republic is not. However, there is a plausible explanation for the Czech response. The Czech Republic has become a tourism 'hot spot' and, as a result of increased numbers of foreigners, commercialism is clearly much more evident than it was a few years ago. As with all such significant changes, even in the most modern, cosmopolitan societies, the local people feel a sense of loss when their favourite coffee shops, restaurants or picnic places are taken over by tourists. This is how we should understand the Czech response.

International trade and influence

Another somewhat similar theme – whether increased international trade and business links were considered a good thing or not – was also explored in the Pew survey. The respondents in all countries had a strong positive response, the highest being Vietnam (98 per cent), then Uzbekistan (97 per cent), Ukraine (93 per cent), China (90 per cent), Bulgaria (89 per cent), Russia (88 per cent), the Slovak Republic (86 per cent), the Czech Republic (84 per cent) and, last, Poland (78 per cent). The results are not surprising. The idea of autarky is no longer popular – if it ever was – except in North Korea and among a few of the anti-globalisation zealots.

A very similar question was whether or not it is a good thing that the world was becoming 'more connected through greater economic trade and faster communication'. The results were not that different from those of the previous question: Vietnam was on top of the list with a very high percentage of its people having positive response (97 per cent), followed by Uzbekistan and the Czech Republic (96 per cent), the Slovak Republic (94 per cent), China (90 per cent), Poland and Russia (89 per cent), Ukraine (87 per cent) and, last, Bulgaria (79 per cent). If the views of the people of these countries are the yardstick, a globalised world is popular and is here to stay.

It is as producers (workers and business people) that we are concerned about globalisation, due to the harsh reality of competing with vastly varying wage rates. When we remove our overalls and go to spend our hard-earned money as consumers we discover that globalisation offers a major benefit with its cheap goods. This is the dilemma.

With regard to the influence of foreign popular culture (movies, television and music), only the responses of the Russians, with 32 per cent, and Uzbeks, with 26 per cent, that the influences were 'bad' stand out. Respondents from all other counties had a high regard for popular foreign culture. Russia and Uzbekistan are special cases as we have noted.

The rich get richer

The survey asked respondents what they thought had happened to the gap between the rich and poor over a five-year period (1998–2002). People from all countries except the Ukraine reported that the gap had increased. Other than Vietnam, with the 'got worse' response at a relatively low 51 per cent, the others responded that the gap had 'got worse' at percentages of 84 per cent and above. This finding is consistent with survey results on this question in numerous other countries around the world – even where the perception of an increasing gap might not mesh with reality.

Perception is reality when it comes to the distribution of the benefits of free markets and increased global trade. The all-too-real widening income gap is the most important reason to oppose modern market societies. With the exception of the social democratic Europeans, the greatest benefits of globalisation are allowed to flow to the rich and, in the case of Russia and a few other chronically corrupt countries, to the criminal class of robber barons.

WHAT HAPPENED TO MY CULTURE?

Did respondents believe their own culture was superior to others? We find that the countries less integrated into world affairs tend to have the strongest belief in the importance of their own culture. This is easy to comprehend.

The percentages of people who believe in the importance (or superiority) of their own culture were high in the less-integrated countries: Uzbekistan (77 per cent), Bulgaria (74 per cent) and China (60 per cent).

If we consider attitudes to home culture by citizens in all of the 44 nations covered in the Pew survey, it is mainly the West European countries (plus the Westward-looking East European countries such as the Czech Republic), Canada and Jordan that don't believe in their own cultural superiority. This suggests (fairly strongly) that the more open a society, the less insular it is. This, of course, is hardly a new insight. But openness is not just a commitment to democratic elections and the rule of law. It requires much more: a willingness to open one's mind to ideas and to ask the Socratic question 'why?'. At the heart of this is the existence of good non-religious schools.

The citizens of the USA, with 60 per cent believing in their own cultural superiority, are more in line with the less-connected peoples of the world, perhaps because of the influence of fundamentalist religion and insular education. This seems a curious fact given the great mixing of cultures that occurs in the USA, yet it helps explain US foreign policy – and much more.

THE VALUE OF OPEN SOCIETIES

In summary, what we learn from the Pew study is that, with one or two exceptions, the surveyed countries are embracing the opening up of their societies. Trade is considered very desirable; ideas from other places are welcome – and much more. Russia and Poland are exceptions in some matters. Russia was pushed, or pulled, into adopting the wrong model, which has taken it backwards and Russia is only now seeking forward-looking, sustainable approaches to its economic system.

The irony is that the formula given to Russia by the advocates of capitalism – change the economy and the institutions of (good) government and a civil society will fall into place afterwards – is pure Marxism. A more realistic model would have the institutions changed in concert with the economy or even have changed the former first. The ideal situation would have been to put the appropriate institutional framework in place before turning to the economy. This could have been done in Russia – and it should have been done.

Just as the transition from colonisation to independence failed many of the new countries because it was not properly thought through and phased in, so has the crash-through campaign in Russia failed to produce what that country's people deserve. The West should have understood why and how the Russian transition would fail. We have found to our dismay that by not preparing the ex-colonies (of Europe) for democracy and economic maturity we are still, unsuccessfully, 30 to 40 years later, attempting to clean up the mess. We only need to look at sub-Saharan Africa and its tragic state. Providing education and institutions that are prerequisite for a modern society would have been much more effective for the peoples of the ex-colonies. It was the very least we could do. The financial 'whiz-kids' who came from the West to remake Soviet Russia had not learned a thing from history.

Poland as a special case

Poland is not the social and economic disaster that Russia is. Its problem, if that is what it is, is a cultural one – deeply religious people are confronted with the Enlightenment values that underpin modern Europe. The French revolution of 1789 broke the rule of the autocracy and the power of the church over the people of France. Following that, in their own time, other religiously conservative countries in Europe, such as Italy, Spain and Greece, experienced the disassociation of state from the church. The important fact is that these countries underwent a social evolution at the same pace as their economic development. This process, I suspect, awaits Poland. Until it breaks the bonds between church and state, it will not be anywhere near as dynamic a country as it deserves to be.

CHINA IS DIFFERENT

China is a completely different story. Who, a generation or even a decade ago, would have thought it likely that China would draw level with, then accelerate past, the USA in economic prosperity? Some did see the writing on the wall well before the popular media made China's economic success a prominent headline – none other than Goldman Sachs, the global financial services firm, has forecast that China will overtake the USA by 2040.

Be mindful that until China develops a large and politically strong and influential middle class, the country's economic development will fall in line with Kuznet's environmental curve. What this means is that the pollution clean up – and it will be massive – comes later. The pollution from China will be beyond belief. We have come to expect this from those who claim to be Marxists.[14] East Germany was probably the worst culprit prior to China. At the end of its day, East Germany was generating on a per capita basis more sulfur dioxide per unit of gross national product (GNP) than any country in the world. As environmental historian John McNeill says, 'billowing smokestacks carried a totemic aura among Marxists'.[15] That time has passed.

THE FUTURE

These countries in transition, because they have broken the chains of the past, are not as much constrained by path-dependence as the rest of us are. And that is reason to expect that they won't make all the mistakes we did. There *is* a sustainable future for these countries if they can embrace a genuine social democratic ideology.

1   The Pew Global Attitudes Project is a worldwide public opinion survey of more than 38 000 people in 44 countries. The data used in this chapter is from the 2003 survey. Sample surveys for each country involve information from 500 to 3000 interviews. The samples from both China and Vietnam were drawn predominately from urban areas and hence there is a bias in their results.

2   Data from the United Nations Development Program, Human Development Index reports 2000, 2003 and 2007/08. The GDP per capita for the years 1975, 1980, 1985 and 1990 are measured in purchasing power parity in $US at 1995. For the year 2005, the measure is the same except the base year is $US2005.

3   The French colonialists were defeated in 1954. The Americans then entered the country on the side of the South Vietnamese and eventually their involvement escalated to full scale war, which did not end until 1975.

4   Data from the United Nations Development Program, Human Development Index report 2007/08.

5   Thomas Noonan, Scandinavians in European Russia, *in The Oxford Illustrated History of the Vikings*, Oxford University Press: Oxford, 1997, p. 134.

6   Gwyn Jones (*A History of the Vikings*, Oxford University Press: Oxford, 1984, pp. 245–6) describes the founding: the local people, Slavs and others 'said to the people of Rus, "Our land is great and rich, but there is no order in it. Come to rule and reign over us". They thus selected three brothers ... The oldest, Rurik, located himself in Novgorod; the second, Šineus, at Beloozero; and the third, Truvor, in Izborsh. On account of these Varangians, the district of Novgorod became known as the land of the Rus'.

7   Mikhail Gorbachev and Daisaku Ikeda, *Moral Lessons of the Twentieth Century: Gorbachev and Ikeda on Buddhism and Communism*, IB Tauris: London, p. 1.

8   Ignacio Ramonet, *Wars of the 21st Century: New Threats, New Fears* (trans. Julie Flanagan), Ocean Press: Melbourne, 2004, p. 106.

9   Vladimir Putin, *The Guardian*, 6 July 2005.

10 Richard Pipes, *Communism: A History*, The Modern Library: New York, 2001, p. 158.

11 Timothy Byrnes, The challenge of pluralism: the Catholic Church in democratic Poland, in Ted Gerard Jelen and Clyde Wilcox, *Religion and Politics in Comparative Perspective: The One, The Few, and The Many*, Cambridge University Press: Cambridge, 2002, p. 34.

12 Byrnes 2002, p. 32.

13 Yet, in the distant past, the 'stans' (Kazakhstan, Kyrgyzstan, Uzbekistan, Tajikistan and Turkmenistan) were on the trade routes between China and the Mediterranean and were therefore open to considerable outside influences.

14 While Russia rushed into privatisation and the market economy, China tiptoed – at least until 1997 (John Micklethwait and Adrian Wooldridge, *The Company: A Short History of a Revolutionary Idea*, Phoenix: London, 2005, p. 123).

15 John McNeill 2000, p. 89

8

NEW ASIAN DRAMAS

The West is going to be reborn in our part of the world through those of us who are here and are going to be the bearers of Western civilization as it was, in its reincarnated form.

Chandra Muzaffar, Malaysian Muslim intellectual

No serious analysis of sustainable development can overlook what is happening in Asia, particularly in China, some parts of South-East Asia and the Indian subcontinent. In this region of the world, the future is rapid economic growth, capable of being measured (if we wanted to) on a day-to-day rather than a yearly basis. The future of much of South Asia (Pakistan, Sri Lanka and Bangladesh) is less certain and it is more problematic.

THE ASIAN TIGERS SHOWED THE WAY

Over the past 40 years, Japan and then the Asian tigers (Singapore, Hong Kong, Korea and Taiwan) have showed what is possible given limited natural resources but a well-trained workforce. Are they the model for the rest of Asia, which represents one-third to a half of the human population?

The idea of Asia – and the term 'Asia' – originated with the Greeks. For them Asia was where the sun rose, hence the definition of Turkey, Iran and Afghanistan as 'western Asia'. Where in the east Asia started, the Greeks did not know. Today we are likely to answer Japan, with far-eastern Russia in the north and Indonesia in the south-east as the other boundaries. In this chapter most attention will be given to two South-East Asian countries, Thailand and Indonesia, which are at different stages of development and with significantly different numbers of people and resources, politics and religions. With all the emphasis on China at present (and the increasing attention being paid to India), it is instructive to consider two countries less in the limelight.

DIVERSE ASIA, ADVANCED ASIA

Asia today comprises diverse religions – Buddhism, Shintoism, Taoism, Confucianism and Christianity[1] – belief systems and cultures. It pays to remember that some areas of Asian civilisation were the most advanced for much of human history. While we can rightly give the Greeks the pride of place for initiating and then developing the philosophy and science we rely on today, we should recognise that throughout the Dark Ages the West regressed while Asia, in particular China, progressed. Table 1 highlights Asian inventiveness and shows that China, in particular, was ahead of Europe in many ways.

Table 1: Dates of innovations in Asia and the West[2]

INNOVATION	ASIA	THE WEST
cotton cloth or clothes	*3rd millennium BC (India)*	*16th century AD*
iron casting	*2nd century BC (China)*	*13th century AD*
planting with automatic seeder	*1st century BC (China)*	*1700 AD*
curved iron plows	*1st century (China)*	*1700 AD*
paper	*2nd century (China)*	*1150 AD*
rotary fan for ventilation	*2nd century (China)*	*1556 AD*
wheelbarrow	*3rd century (China)*	*1200 AD*
porcelain	*3rd century (China)*	*1709 AD*
watertight compartments in ships	*5th century (China)*	*1790 AD*
printing with wood blocks	*8th century (China)*	*1400 AD*
gunpowder	*10th century (China)*	*13th century*
bombs	*1000 AD (China)*	*16th century*
printing with movable type	*1045 (China)*	*1440*
iron-clad warships	*1592 (Korea)*	*1862*

THE GIANT – CHINA

Throughout history China has been the giant in Asia, not only in its own achievements but also in terms of the influence of the Chinese diaspora throughout much of Asia. Therefore, a brief discussion about China will help set the scene for what follows. China is, if all goes according to realistic projections, on target to join the USA and the EU as the world's third superpower. The UNDP's human development index (HDI) of 2005 presents the following picture of China. It is ranked number 81 out of 177 countries with a gross domestic product per capita (PPP) of US$6757. Over 75 per cent of the population has access to safe water, with 44 per cent having access to improved sanitation. Using conventional GDP indicators, the rate of growth in China is nearly 9 per cent per year. At that rate, per capita income doubles every seven years.[3] Let us briefly look back at China's history, so as to underscore its position in the world today.

A brief history

Classical China was sophisticated well before Rome was. Chinese administration, undertaken by highly trained mandarins (we have coined the term 'literati' to describe the modern equivalent of this class), was legendary. Not only did the Chinese have home-grown expertise, but also they learnt from abroad. The Chinese were inveterate travellers and settlers in foreign lands. For millennia they travelled, over land rather than sea. They moved into South and South-East Asia, where large Chinese populations remain today. Then in 1402, Emperor Yongle came to the throne and China became a major maritime power. He instructed Zheng He to mount a series of maritime expeditions.

Zheng He's first journey set forth at the beginning of the fifteenth century for what is now Indonesia (to Java and Sumatra), passing through the straits of Malacca, then to Sri Lanka and India before returning to China. He had 62 ships – ships, not boats! At the end of that century, some 90 years after Zheng He set sail, Columbus made his first trip to the Americas with three boats, the largest of which, the flagship *Santa Maria*, was about 26 metres long. He had a crew of 90 men and boys.

Zheng He made the last of his seven expeditions from 1433 to 1435. He sailed from China to Java, turning west to Arabia and then travelling down the east coast of Africa, before heading back home to China. On this voyage he had under his command 317 ships with a total crew of 27 750 men. The largest ship was about the length of a modern destroyer, approximately 135 metres. No sooner had he completed his journey than a palace revolution occurred back home, and with this the attitude of the Chinese to the rest of the world changed. Foreign trade and travel were taken off the political agenda and it has stayed that way until very recently.

Prosperous China

Even though China turned its back on the world then, it retained the relatively high standard of living it had achieved during its period of maritime adventures and foreign trade until the eighteenth century. The journalist Nicholas Kristof informs us that the 'average Chinese lived longer in the eighteenth century, ate better, and enjoyed more luxury goods than a European of that period. For example, China consumed almost five pounds of sugar per capita in 1750, compared to two pounds in Europe. Another luxury, tobacco was much more common in China, so much so that even toddlers smoked'.[4]

Missing the boat

Given these facts, and others that provide additional evidence of China's success in the not-too-distant past, we can join with Kristof in asking what went wrong. We should not concur with those who assert that the Chinese lacked a passion for business – to the contrary. Maybe it was the consequence of nothing more than a strong leader closing the door to trade just as the world was opening up. This was both metaphorically and literally speaking a case of 'missing the boat'. A poor call!

Communist billionaires galore

Present-day China is as difficult to fathom as it was during periods in the past. China is an enigma, a so-called 'communist' country with probably (as it is difficult to obtain reliable data) more billionaires than the USA. It is only just over a generation ago that Mao Zedong ruled China, with the help of millions of Red Guard cadres, on the formula that the difference in income between the poorest and richest citizen should be no more than fourfold. The conventional wisdom was that Mao's salary, as the most powerful person in the country, did not violate this rule.

Anyone with the slightest interest in making a profit was 'counter-revolutionary' or 'bourgeois' – whatever the vilest description happened to be at the time. These same people, or their sons and (much fewer) daughters, are now officially 'advanced productive forces' in the official jargon or, in plain language, successful capitalists. It is worth keeping in mind that China's start on the road to capitalism pre-dates by a decade the collapse of the Soviet empire and its unsteady attempt to alter the nature of its economy. The Chinese acceptance of capitalism was deliberately designed and finely managed, rather than haphazard, as in the case of Russia. This helps explain the more constant and vigorous growth in China. The fact that the big stick of the communist party can be utilised – when necessary – to assist profit-making also distinguishes China from Russia. Both economic and political power disintegrated when the Soviet Union collapsed, but this was not the case in China.

There is much about modern China that is impressive – and much that remains depressing. The hope of most observers is that some sort of progressive, liberal political and social transition will build upon the change of the economic base. This is how modern China could march in tune with Marx's 'Marxism' rather than that of revisionists of today. Recall that Marx saw the economic base of a country being the force that fashioned its social, political and cultural norms. The new capitalist China urgently needs the institutions that give capitalism a human face in Europe, Australia, Canada, New Zealand, the USA and, eventually, I hope, China may embrace the social democratic institutions of Scandinavia.

China today

Today China is the world-leading producer of grains, meat, cotton cloth, textiles, television sets, steel and cement. It is also a leading coal producer, the fuel that drives the other sectors of its economy. However, it still needs to import vast quantities of coal to fuel its modernisation. China is number two on the world scorecard for chemical fertiliser production and application (to feed its grain industry, which feeds its meat production, which feeds its human population) and number three for electricity production. The country is the world's leading consumer of meats and grains, as well as fertilisers, steel and coal.[5]

Much muck

China, as they say, is very much 'a work in progress'. And the progress is so rapid that what is written today is out of date if not tomorrow, then the day after. Its cities are blighted with pollution. And much more muck is to come. In 2005, a Chinese energy firm, CNOOC, attempted to purchase the US energy company Unocal. The bid was frustrated by the US government – and the US president. The chief financial officer of CNOOC, Yang Hua, was extremely annoyed and frustrated. He is reported to have said that to deny China access to energy amounted to a violation of human rights. He asked, 'What are human rights?', answering emphatically and dogmatically that it means having petroleum to run your car. We cannot deny the logic of this, particularly from a materialist perspective. However, the Chinese car-owning class, plus the rest of the world's car-owning population, are going to need to find a way to deal with climate change and the other environmental consequences of the private car economy.

How far into the future will it be before China has the desire and necessary surplus – which will need to be large and in government hands – to start cleaning up the pollution? How much damage to human health and the environment will occur? Much is written about helping China develop without making the environmental mistakes of the West. Headlines announcing massive aid dollars and technology transfer appear regularly; however, the evidence that this is making a difference is nowhere yet to be seen. To date, there has been much talk and too little action. How China deals with its environmental problems in the next decade will be one key to the future of the world.

SHADOW PLAYS – INDONESIA

A very mixed culture

We leave China and turn to our other chosen countries, starting with Indonesia. The Malay people commenced to settle in what is now Indonesia some time around 2000 BC. There were, as we know with the discovery of 'Java man' in 1891, human predecessors a million years before then, but here we will stick with the more recent past. After the Malay people, the Indonesian islands were colonised by Hindu and Buddhist peoples of southern India, seeking gold and spices – and converts to their religions. The island of Bali, with its endless colourful Hindu festivals, is a legacy of Indian colonisation. The Indian advance to Indonesia came to a halt when the Indian leaders' attention was diverted to domestic matters by the Muslim invasion of northern India during the fifteenth and sixteenth centuries. A small irony! The Chinese settled the Indonesian islands in the ninth century. The Chinese, as we have come to recognise, were excellent traders and they focused on the islands of Java, Sumatra, Timor and Sulawesi. Among a wide range of goods they sought were trepang (bêche-de-mer) and birds of paradise – both extremely valuable items in China.

It could be expected that the various colonisers of Indonesia would have had a significant impact on its culture, and so they did. Yet the new cultures, and particularly the religions, had to build on or at least accommodate the existing animist cultures and belief systems. Even today, religion in Indonesia – conventionally recognised as a Muslim country – is a fascinating and, at times, weird blend of ancient and modern supernatural beliefs. Except for a small clique of fundamentalist Muslims, Indonesia is the least Muslim of the Muslim countries (I don't regard Turkey as a Muslim country).

The ancient past, including the mysticism it entailed, is still a part of Indonesian culture. The Indonesians are a cultured and aesthetic people, with their *wayang* shadow plays and *gamelan* orchestras of gongs and xylophones, as well as their carving and painting. The meeting of the old and the new – the new being the coming of Islam only a matter of some hundreds of years ago – helps explain why Indonesian Islam is significantly different to the Islam of the Middle East. For the most part, the cultural and religious mixing in Indonesia has led to a moderate, easygoing Islam. This is not to dismiss the small cells of fundamentalist Muslims that exist in parts of Indonesia. Circumstances conducive to violence and a mix of Islam and ancient myths can bring horrific results, as the story in Box 1 illustrates. If nothing else, this story proves that belief in the supernatural is extremely dangerous.

> **Box 1:**
> **The Asian financial crisis excites the ninja**
>
> Journalists Nicholas Kristof and his partner, Sheryl WuDunn, describe in their 2000 book *Thunder from the East: Portrait of a Rising Asia* their travels to Java to study the impact of the Asian financial crisis on rural Indonesia. In the eastern Javanese town of Mojokerto, in a conversation with a rickshaw driver named Salamet, Kristof is informed that the Javanese have always practised black magic and sorcery. The sorcerers are known as 'ninja'.
>
> Salamet explains how, as the 1997 financial crisis took effect, sorcerers took advantage of the confusion and that black magic began to used again. He makes reference to 200 murders, mostly of Muslim leaders, recounting how bodies were chopped into pieces and hung in trees. Then

came the reciprocal killings, whereby suspected witches and sorcerers were stabbed to death or beheaded, then their bodies dragged through the town's streets in a display of might. This was violence far more cruel and crazy than that which I had experienced in the mid 1980s in Java when riots targeting Chinese business resulted in some dozens of deaths.

Salamet goes on to explain to Kristof that the Muslim youths who took it upon themselves to slaughter the ninja believed that if one ninja was killed 1000 more would replace him or her (ninja came in both sexes). While we would ask that if they multiply if killed why kill one, the myth-believing youths didn't think to ask this question. There is no rationality when extreme faith takes over.

The Muslim youths also believe that the ninja are capable of turning into cats! Given this morbid story and the hocus-pocus that drove this insanity, Kristof is understandably pessimistic about the future of Asia. He writes: 'Asia may hobble back eventually, but when people start running around hacking off each other's heads, they are not on the brink of a middle-class or an industrial revolution.'[6]

This assessment, though, is not fair to the middle-class urban Indonesians who sip caffe latte in modern air-conditioned coffee shops while discussing the price of oil (the increasing use of this commodity is, on one-hand, a boon for the oil-rich Indonesia and, on the other, an added expense for its car owners).

A little geography

Indonesia is a fascinating country in many respects. Consider some of its demographic, geographic and societal characteristics. The small island of Java is the most heavily populated place on the earth, while, on the other hand, there are Indonesian islands that are, for the most part, virgin rainforests, having experienced minimal human impact. Only a generation ago headhunters lived in these jungles. Indonesia has the fourth largest population in the world but most of it is concentrated in Java and a few major cities.

Within an hour or two by boat ride from stinking, polluted Jakarta Bay is the so-called Thousand Islands marine playground – if not a thousand islands, there are a very large number of small coral islands and coral reefs. Where they have been protected from crude fishing methods (dynamiting and poisoning) and exploitation of limestone for cement making, these reefs are close to being serious challengers to Australia's Great Barrier Reef in beauty and diversity, as are reefs off the shore of some other Indonesian islands.

Due to its size (a land area of 182 644 000 hectares) and geography, approximately one-half of Indonesia is covered by tropical forests. Of these, 21 per cent is protected. The country has 2 390 000 hectares of mangroves, with a third protected. While these numbers are impressive, population pressure on Java, forestry operations in Sumatra and Kalimantan, and coastal development in general are serious threats to the country's biodiversity. Already 15 species of primates, 6 species of big cats, 8 species of ungulates, 46 species of rodents, 32 species of bats, 19 species of reptiles and 113 species of birds are threatened.

A little sociology

Along with its incredible diverse ecosystems, there is great diversity in Indonesian culture. Only by creating a new language, Bahasa Indonesia, was the postcolonial Indonesian government able to form a nation; the existing enormous range of languages and cultures had to be leapfrogged. It is less difficult to fight over something completely new than over attempts to build on existing languages. The alternative – to wait for a natural cultural blending – would have required far too much time and the nationhood would still be evolving today. There was a degree of urgency to cutting off the past to build an economically and politically viable community – and, amazingly, it worked. Compare it to the former Yugoslavia, which fell into religious conflict as soon as its strong leader died and of which various regions are still fighting – if no longer in the streets, in their minds.

Today, the Indonesian nation is searching to overcome a resurgent religious and ethnic divide. An apparently successful multicultural society could too readily collapse if this new threat is not curtailed. The Indonesian problem is no different from that in other moderate Islamic countries where fundamentalists have been able to establish a foothold because moderation requires forgiveness of differences. This is the most basic problem of liberalism.

Until recently, visitors to Indonesia would not know that they were in a Muslim country – just as if visiting Australia they would not know they were in a Christian country, by inclination if not by prescription. The secular nature of Indonesian society is the gripe and the target of its increasing number of Muslim fundamentalists. The headscarf has reappeared in recent years; the first sign of the oppression of women according to some Muslim modernisers. One can only wonder what type of society Indonesia would be today had Sukarno, the first president of the newly independent country, stayed on as leader past 1967, and had the Indonesian Communist Party been allowed its democratic right to contest elections. Rather than China being the dynamic economic force in Asia, it is possible that Indonesia would have set the pace. Indonesia has been robbed of its future by major corruption at the highest level and a stagnant society has opened up the ground for extremists.

Jakarta, probably more than any other Asian city, represents the old and new, poor and rich Asia. Until it was burnt down in the mid 1980s, the largest supermarket in town, Sarinah (named after one of Sukarno's wives), was only a few storeys tall and was easily dwarfed by the average suburban shopping complex in Singapore or Australia. Only in the past 20 years have the high-rise international hotels and the ubiquitous golf courses proliferated. They sit alongside the densely populated urban villages (*kampongs*) and some of the world's worst slums that are the real face of Jakarta for a sizeable percentage of its population.

We despair for the human race when we see mothers with young babies living in cardboard shelters under the bridges that cross some of the world's most polluted and stinking canals. I'm writing of Jakarta today. The international visitor or visiting foreign expert, living in a five star hotel and driven in a chauffeured car, is likely to miss the real Jakarta. The canals (of which fourteen slice through the city) are public bathhouses for some and toilets for others.[7] Desperate people make a few rupiah a week by scooping out of the slow-flowing canal water and muck the debris of the consumer society – plastic bags and bottles – to be sold for a pittance to recycling companies or to unscrupulous traders who refill the bottles with soft drinks. The wretched canal scoopers are doing their bit for the environment – through necessity.

The dichotomy of development

We do nothing for these poverty-stricken people. Within 100 metres of the pathetic struggle for life I have just described will be a McDonalds or KFC and an air-conditioned shopping centre displaying Paris fashions, Italian clothing and Swedish furniture. The young Indonesian 'yuppies', with European-made mobile phones conspicuously displayed, drive there in their Japanese cars or, if they are richer, in their BMW or Mercedes.

The trickle down of money to the poor will, ever so slowly, occur; however, over the many decades it takes to make a real difference for the poor, how much suffering will occur? The poor I met in Indonesia nearly 30 years ago are likely to be dead – even though then they would have been much younger than I was. Those who could do something about this human tragedy, don't. They don't want to know and some don't care. The slum dwellers of Jakarta are the equivalent of the factory fodder of the nineteenth-century Industrial Revolution in England. Do the poor of Asia have to follow the same path? Maybe two or more generations will have to pass before living wages are paid, before working hours are in line with those of the developed world, before a social welfare system that would meet nineteenth-century Enlightenment principles is in place.

Life for the more fortunate Indonesians can still be difficult, as shown in Box 2. A snapshot of the lives of a small group of Indonesians – from poor to middle class – based on interviews recorded in the streets of villages and cities of Indonesia in early 2004 highlights the issues facing modern Indonesia.

Box 2:
From the bottom up

Mat Sai comes from Malang. He holds down three jobs for a monthly income of around US$18. His wife works as a contract labourer in a cigarette factory. He lives in his mother's house together with his mother and younger brother, wife and daughter. The high and unstable cost of essential items is the most significant problem facing him. He argues that poor families should be able to meet their basic needs and their children obtain a good education; however, he is pessimistic about Indonesia's future. He believes that the necessary reforms will be difficult, as good government requires that the electors are informed and desire change. This means, in his words, that 'people from the bottom up have to change'.

Agus is 32-year-old street-stall vendor with his own stall on the edge of a street in Jakarta. Agus moved to Jakarta with his wife and children in the hope to find work in the 'big city', but has found it difficult. He works every day for 14 hours per day and earns enough to support his family. He feels that Indonesia's biggest challenge is the economy and at the heart of its issues is corruption and lack of safety and security.

Nurhayati and her husband own a small shop, selling household goods and medicine, in a traditional market in Jakarta. They open the store every day and, together, they earn about US$70 per week, which they use to support themselves and their parents, an amount she feels is enough. She believes that Indonesia faces multidimensional economic, political and moral problems, but she feels that, with honest leaders, many of the problems currently experienced by Indonesia could be resolved.

> Iin is a 22-year-old live-in maid in a residential house in Tangerang. She completed elementary school, but could not continue her schooling for financial reasons. She works seven days a week for 12 hours every day and earns US$20 per month. She is content with her life, but would like a higher salary so that she can send her child to school, as she currently is not able to afford this.
>
> Iin believes the poor state of the economy is the most significant issue facing the country, a result of government corruption in the past and a lack of law enforcement. She asserts that that if the government was committed to the elimination of corruption, Indonesia could become a more secure country.

There might be more fascinating cities than Jakarta, but they are hard to find. The Soviet-style statues saluting working-class heroes and soldiers – which you can't miss on your taxi ride from the airport – are, as John Hughes writes, 'sadly alien to so cultured and artistic a people as the Indonesians'.[8] The irony is that the Indonesian government has not celebrated working-class heroes since the fall of Sukarno.

The smells from the roadside vendors with their delicious, chilli-garnished but possibly diarrhoea-inducing snacks, the ever-present *kretek* (clove) cigarette smoke, the grit in the eyes and taste of fumes in the mouth are a continual reminder that this is Asia. Not the Asia of Singapore, with its chewing-gum-free public transport and sanitised chic shopping complexes; not the Asia of Hong Kong, with its buildings reaching to the heavens. Of course, Jakarta, and Indonesia, are not Tokyo or Japan.[9]

The Mad-Max traffic – still dominated by *bajajs* (the filthy, but economic because pollution in not counted as a cost, three-wheeled transport), Russian trucks and buses (which now must be 50 years old and hence have a propensity to break down in the main urban street, Jalan Thamrin), motorcycles in every available square metre of road space and increasing numbers of air-conditioned Japanese sedans – is a reminder that Jakarta is Asia somewhere between environmental catastrophe and significant wealth.

How slow is change?

Read Hughes' 1967 book, *The End of Sukarno: A Coup that Misfired, a Purge that Ran Wild,* and be struck by how little things have changed in over 40 years. Hughes sets the scene for the 'night of the long knives', 30 September 1965, when the abortive coup against Sukarno took place:

> *As the day's heat ebbed ... the day's gasoline fumes faded, the liberated evening air took on a gentle fragrance peculiar to Indonesia, an exotic blend of jasmine, frangipani, and the smoke from locally made cigarettes spiced with cloves ... The city begins to unclog and untangle from its daylong traffic smell ... eighty percent of the city's buses are off the streets for lack of spare parts. Those remaining are groaning, overloaded testimony both to the skill of their makers and the ingenuity of Indonesian mechanics who keep them stuttering along on bent wire, and metal salvaged from beer cans. By the nightfall this tatterdemalion old fleet has coughed, spluttered and belched its way homeward, laden to the long suffering gills with Indonesian workers ... Then the streets are left to the tattered bejak-drivers ... And as the bejak-drivers clang their deep gongs, and the little kerosene lanterns flare up on street-corner stalls ... Djakarta at night seems touched a little by the magic of Java, the strange and mystical island on which it is set, the headland of the 3000-mile-long Indonesian archipelago.*[10]

For those interested in the 1965 coup, a fundamental turning point in modern Indonesian history, see Box 3.

Box 3:
The October 1965 coup and the frenzy that followed

We still do not know beyond a shadow of doubt who planned the coup against the president in 1965 or why it was attempted. Some things are clear. The communists (the PKI) were gathering support and becoming a significant influence on Sukarno and deputy premier and foreign minister Subandrio. Were these two moving too far to the left in the eyes of the armed forces, business and their foreign supporters?

Over the years there have been charges of CIA involvement. John Pilger, for one, believes this to be the case, while Hughes comments, 'If the CIA was involved in this operation, it must be the most brilliantly disguised secret in the entire history of the agency'.[11] Who is right?

The response to the attempted coup was one of the greatest bloodbaths of the modern era. Hughes called it 'the great crackdown' but there are only vague estimates of the numbers killed. The Indonesian minister for foreign affairs in 1966, Adam Malik, is reported to have said, 'We'd never taken a census before the coup. We didn't take one after'.[12] *The Economist* suggested the number killed to be one million people, while the *Washington Post, Life* magazine and the *New York Times* reported it was more like half a million. The vast majority were communists, their families, people rumoured to be communists and Chinese-Indonesians, of whom there were between two to three million at the time. Many Chinese-Indonesians were dominant in the economy because of their business skills and their success was resented. This was the supreme irony

– successful capitalists were murdered in the cause of eliminating communism.

Suharto and his army generals played the race and religious cards when they gained control.[13] Muslim youths were encouraged and given free reign to slaughter communists and suspected sympathisers, in many cases whole villages (including women and children) that were suspected of harbouring communists.

Not all children were innocent, as Hughes explains: 'Whole villages, including children, took part in an island-wide witch-hunt for communists, who were slashed and clubbed and chopped to death by communal consent'.[14] In Bali the frenzied mobs weren't Muslim but rather Hindu. Hughes quotes a Balinese person:

This may be difficult for the Westerner to understand. But what happened here was a sort of mystical cleansing of all the island's problems and ills. Things had not been going well for years. Balinese had to labor under the Dutch ... then under the Japanese ... then came the revolution to make Indonesia independent, which meant further upheaval. Some years ago we suffered tragic loss of life ... when our big volcano erupted. Then ... we got the communists stirring up trouble. In many people's minds all their troubles blurred into one sense of discordancy. And by ridding the island of communists, they believed that all other problems would somehow be removed ... It was a kind of purifying of the land from evil.[15]

Meanwhile, in the streets of Jakarta, Muslim youth demonstrated in front of the US embassy shouting 'long live America'!

Transition to where?

The juxtaposition of the electronic revolution at checkout counters in world-class supermarkets with one of the most cumbersome bureaucracies in the world and Soviet-realism-style statutes – 'strident as they break their chains and cry death to the imperialists' – says this is Asia in transition. However, it is not obvious that the future of Jakarta is to be another Singapore, or Indonesia another Japan. Indonesia's start on the road to industrialisation and a market economy preceded that of China, yet it seriously lags behind today, notwithstanding the fact that its middle class had grown to about 20 million (a tenth of its population) prior to the setback of the late 1990s financial meltdown.

Financial crisis

The Indonesian people, its poor and middle class, were the real losers in the 1997 Asian financial crisis. As far as we know – and we know little about the doings of the Indonesian political and economic elite headed by past-president Suharto – the powerful, wealthy 'crony capitalists' came through the crisis unscathed. Unlike Malaysia, Indonesia had no Dr Mahathir to say 'no' to the Washington Consensus. Maybe those Indonesians who could have acted as the Malaysian leader did (in stopping the exchange of the Malaysian currency) believed their country (for some reason) deserved to be stripped of its economic wealth and hence did nothing to halt the crazy devaluation. Or perhaps they felt that they could not challenge the likes of the Inational Monetary Fund (IMF) and the World Bank. Maybe those who could have moved to avert the crisis knew they would not suffer from it.

Corrupsi, the Bahasa term for corruption, was rampant and clearly Indonesia's financial infrastructure was not up to standard, yet its natural and human resource base was unquestionably sound. No-one yet has explained what every Indonesian high-school student wants to know: why could he or she buy a packet of sweets for 1000 rupiah one day in 1997 and the next day have to spend 3000 rupiah for the same packet? No wonder that Indonesian students look bewildered when told that economics is the science of rational decision-making.

THE LAND OF THE FREE – THAILAND

Modern Thailand was known as Siam until 11 May 1939 when its name was changed to 'Prathet Thai' – Thailand in English. It means 'land of the free' – an appropriate name for a country that has never been colonised, a unique position in South-East Asia. This is a point the modern Thais will make with a sense of pride if you query them about their history. The Thais' attitude is notable for the confidence and self-assurance with which they conduct themselves in world affairs.

Imagine Thailand as being located at the apex of South-East Asia: in the north it borders Laos, on the north-west Burma, on the east Laos and Cambodia and on the south, Malaysia. China and Vietnam are not far away. The ancestry of modern Thai people is subject to ongoing archaeological and anthropological investigation; however, it is clear that the original – and more recent – settlers came from southern China.

The population of Thailand is well over 60 million, with over 9 million living in Bangkok, according to official counts.[16] Thailand is ranked 78 out of 177 countries in the most recent human development index (HDI). The vast bulk of the population is Buddhist – in name if not always in practice. Muslims, particularly in the south (in the former independent Kingdom of Pattani), are a rebellious minority and their leaders are seeking separation using terrorist tactics. The other minority religious groups, Christians and Hindus, are well integrated into the Thai nation.

The economy

Thailand's major economic sectors consist of tourism, textiles and garment manufacture, light manufacturing such as electric appliances and parts, furniture, plastics, forestry products, seafood and various crops, of which rice is paramount. Thailand still has a very large agricultural sector (about 80 per cent of the population works in agriculture), with numerous small farms (on average 10 *rai* or less than 1.6 hectares). As with much of South-East Asia, Thailand is experiencing a significant drift of the rural population to the major cities, in particular, to Bangkok. This can be managed while the industrialised and construction sectors continue to expand but when things go wrong in these sectors, both the urban and rural economies suffer, but in different ways as we will see.

The financial crisis

Prior to 1996, the consensus was that Thailand would be the next 'newly industrialised country', following South Korea, Taiwan, Hong Kong and Singapore. Then the unexpected happened. The Asian financial crisis from 1996 to 1998 saw vast numbers of poor workers lose their jobs in Bangkok and other large cities and, as a result, drift back to their rural roots. The knock-on effect as the economy slowed rather dramatically was that middle-class workers also lost their jobs. The return of workers to rural areas threatened the sustainability of the farming community. There was little the Thai government could do but it responded by initiating the king's idea of the 'sufficiency economy' and his 'new economic theory' in rural areas of the country where small-scale farmers dominated.

There are various goals of the king's economic rubric, such as self-sufficiency in rice production, mixed farming (growing crops other than rice), growing fruit trees, raising chickens (or a small number of farm animals), pond aquaculture and self-sufficiency in water (particularly in drier areas of the country). These on-farm innovations have the objective of broadening the farmer's income base leading to overall self-sufficiency. Once this is achieved, farmers are encouraged to form local cooperatives to market cash crops (eliminating the middleman) and to purchase discounted production inputs such as fertilisers. At the village level, the 'sufficiency economy' is to be built on the 'one village, one product' concept, where village funds are to be used to make the most of the competitive advantage a village community might have, or might develop, to produce a marketable product. Travel through rural Thailand and you'll soon discover villages specialising in products such as colourfully painted wooden ducks, a popular purchase for tourists. Is this a workable solution for a country in its development phase?

Underpinning this is the Buddhist philosophy of 'all things in moderation', which in reality would suggest to these small-scale, relatively poor farmers that they should be satisfied if they can achieve the modest benefits this philosophy sets out to deliver in economic terms. The objective, supported both by the highly revered king and the government bureaucracy, is that that people stay on their land rather than be tempted by the bright lights and the dim possibility of riches in the city. The financial crisis illustrated what could go wrong.

This experiment in rural policy needs to be watched. If it can deliver the same value of produce per hectare that large-scale commercial farming can, it promises to be a solution for any developing country with plentiful farmland and a large rural population – something many of them have. Conventional wisdom suggests that a sufficiency economy should be a thing of the past, that it forgoes the economies of scale of large-scale agriculture and that it does nothing but condemn another generation of Thai farmers to subsistence living. On the other hand, imagine the economic and social problems that would beset all the major Asian countries (China, India, Pakistan, Bangladesh, Indonesia, Vietnam, Philippines) if large-scale market-based farming was put in place overnight. Only a slow transition is feasible and 'small is beautiful' might still be the best policy at this stage of development.

The inevitable environmental problems

In addition to the challenge of helping the rural poor, Thailand faces the usual array of environmental problems we have come to expect in a developing country. Water, as is the case everywhere in the world, is required in certain quantities and qualities at particular times in specific locations. As the old adage goes, it never rains but it pours. Thailand suffers from periodic floods. The flooding of Bangkok during torrential downpours in the wet season is a regular, and widely reported, phenomenon – flooded is probably one of the better known images of Bangkok.

Thailand has a few lengthy and impressive rivers; for example, the Chao Praya, which flows from northern rainforest mountains to the sea near Bangkok. Irrigated rice growing and hydroelectricity are the benefits of a major dam on this river, but sedimentation and siltation caused by slash-and-burn agriculture on the steep slopes is a problem. Some claim the regular flooding of Bangkok is one of the serious consequences of this.

In urban Thailand, water pollution remains a serious problem, notwithstanding the fact that 87 wastewater treatment plants have been built. According to the HDI, 99 per cent of Thais have access to improved sanitation and clean drinking water. These are extremely good results for a developing country – but there are still some issues with water quality. The allocation of the water resources of the Mekong River is, and will be, a difficult trans-boundary economic and political issue as the river forms the border between Thailand, Laos and Cambodia.

The decline of natural resources

Thailand's natural resources – in particular, its rainforests, mangrove forests, coral reefs and wild-capture fisheries – to varying degrees have been depleted. The country has a total land area of 51 400 000 hectares of which 16 237 000 hectares are tropical forests – about one-third of this area is protected. There are also 36 000 hectares of temperate forests, with 11 per cent protected. Mangroves cover 509 000 hectares, but only five per cent are protected. Numerous animals are under threat: 4 species of primates, 3 species of big cats, 8 species of ungulates, 5 species of rodents, 4 species of bats, 16 species of reptiles and 37 species of birds.

Using dynamite and cyanide to harvest fish has destroyed parts of the country's magnificent coral reefs. The untreated or poorly treated waste from coastal tourism resorts has further damaged the marine environment, which is crucial for much of the country's extremely successful tourism industry. While yields from marine fisheries have increased recently, the catch is beyond long-term sustainable limits. The harvest of freshwater fish has experienced a significant decline. Aquaculture keeps expanding, exporting product to a fish-hungry world and these days causing less localised environmental problems that in the past, yet a question mark about its sustainability remains.

Cities gasping for oxygen

City life, particularly in Bangkok but increasingly in Chiang Mai and other larger cities, requires living with environmental pollution that is among the worst in the world. There are days in Bangkok when better-off residents purchase oxygen from drop-in cafes. The street air contains far too much carbon monoxide for healthy living. Exacerbating this problem is that, notwithstanding vastly improved public transport (the skyrail in particular) and good quality freeways, the city comes to a standstill in peak hours with vehicles aimlessly idling, adding to the pollution while burning increasingly scarce fossil fuels. The last point is lost on the Thais as the government subsidises fuel. It is not just the air pollution that causes problems. Noise exceeds the accepted level of 70 decibels at busy intersections. The canals still flood, as noted above, and otherwise stink. Much remains to be done before Thailand reaches the status of a 'newly industrialised country'. It will be though, according to most pundits, the next South-East Asian country to take this step.

The Italy of Asia

Thai politics remains unpredictable, just like Italian politics. A bloodless military coup on 23 February 1991 was, nevertheless, a real surprise. It was only one of the ten successful coups since 1932, of which only two, including the one in 1991, overthrew a democratically elected government. In May 1992, there were large demonstrations – mainly involving students – in which protestors and the military confronted each other near the democracy monument in Bangkok. About 50 people died and hundreds were injured. An election later that year restored a semblance of sanity. Since then the constitution has been rewritten – but just when we thought it was safe to call Thailand a democracy, another military takeover occurred in 2006 and, in 2008, protestors again took to the streets against the government. The government changed hands. Life goes on in Thailand.

Modern Thais

To what extent have the economic and political changes that have occurred in the last 30-odd years impacted on Thai culture at the family and village level? A large middle class, particularly in the major urban centres and tourist towns, has developed and there is much consumption of the goods and services we would find in Singapore or Sydney.

Recall Thailand is a Buddhist nation, where the notion of 'the middle way' – no extremes in anything we do or aspire to – is the central tenet. It is a form of the 'golden rule',[17] which appears in all major religions and philosophies. We also note that traditional Thai society is hierarchical and that this is not confined to economic class relationships but also applies to roles in society and family. One of the positive social traits in such a society is respect and even obedience to elders. To what extent is this changing? It is hard to know.

Reciprocal giving of gifts is another feature of Thai society. Some argue that, as a consequence of these qualities, there is far less emphasis on the individual and more on community than there is in Western societies. Showing respect and not being controversial or argumentative is part of normal life in Thailand – hence 'the land of smiles' and the politeness. Yet, pick up a Thai tabloid and scan the crime pages for another picture of modern Thailand – there is a fascination with gory photographs of victims and demeaning ones of crime suspects. Whether this is any less uplifting than the 'page three girls' of Western tabloids is debatable, but it does reinforce the point that Thai society is complex.

The future?

So what of the future of Thailand? Thailand did recover fairly rapidly from the Asian financial crisis and today has been restored to its pre-crisis standard of living – in fact, it has moved ahead in certain sectors. For many Thai citizens, recent years have brought about significant improvements to their lives. This is shown in the stories in Box 4, which are based on interviews undertaken in 2004 across Thailand. Do they expect, even in the current times of global and local unrest, that these good times will continue?

Box 4:
A better place to live

General Chumphon Jindaghot is a healthy 74 year old, having spent 38 years in the service of the Thai government as a soldier. He has travelled to places that only the better-off Thais (or students who win scholarships) can afford: France, Italy and China. National stability is his major concern for Thailand and, for the world, it is peace. Looking back over the past five to ten years, he firmly believes that Thailand has become a better place to live due to its economic growth.

Terd Hanchai has been a suburban scavenger for three years. His occupation is quite different from that of the scavengers (or 'rag pickers') that live in the large municipal waste dumps of cities like Jakarta and Mexico City. Terd is more of a 'Steptoe', a suburban waste collector and recycler. Previously, he was a manual worker but his present job provides a better living. He is 45 years old and has two children and today his major concerns are about his family and the education of his children. Although he acknowledges Thailand's recent economic growth, he believes that, overall, things have become worse due to the increased cost of living. It is no surprise that, in his view, the major issue facing the world is to clean up the environment – of course, if it did, he would be out of a job.

Atisuda Na Nakom has been a public servant for two years and before that she was a news reporter. She is aged 30, single and highly educated. She worries about the unity of her country and the Muslim separatists in the south. She believes that the improvement in infrastructure in the past five years has been good for the people, while the downside over the past ten years has been the degradation of natural resources and the increased social problems.

1   It is commonly stated that half of the South Korean population is Christian.

2   Nicholas Kristof and Sheryl WuDunn, *Thunder from the East: Portrait of a Rising Asia*, Alfred A Knopf: New York, 2000, p. 29.

3   It took over 30 years for Japan to double its per capita income in its strong growth period. The Asian tigers achieved a doubling of per capita income more quickly; for example, South Korea took eleven years.

4   Kristof and WuDunn 2000, p. 40.

5   Norman Myers and Jennifer Kent, *The New Consumers: The Influence of Affluence in the Environment*, Island Press: Washington, 2004, p. 67.

6   Kristof and WuDunn 2000, p. 11.

7   This was how John Hughes (*The End of Sukarno: A Coup that Misfired, a Purge that Ran Wild*, Archipelago Press: Singapore, 2002, p. 16) described Jakarta's canals in 1967. They served the same purposes when I first went there in 1981, some years later, and still do.

8   Hughes 2002, p. 17.

9   Do we still think of Japan as Asia? This is a serious question – is Japan not more Western than the West? Various radical Asian thinkers assert that Japan has lost it Asianness.

10 Hughes 2002, pp. 15–17.

11 Hughes 2002, p. 113.

12 Adam Malik, quote in Hughes 2002, p. 193.

13 Hughes 2002, pp. 151, 172.

14 Hughes 2002, p. 184.

15 Hughes 2002, pp. 145, 185–6.

16 The unofficial estimate of the population puts it much higher.

17 Do unto others as you would have them do to you.

9

THE VIKINGS TODAY: LEADING THE WAY

Norway's remarkable story of almost continuous surpluses is explained by its prudence in salting away for the future part of the proceeds from its exploitation of North Sea oil deposits.

Rodney Tiffen and Ross Gittins[1]

In the years 2001 to 2006 inclusive, Norway was acclaimed by the UNDP as the number one country in the world on the human development index (HDI). In 2007, Iceland just beat Norway into first place. When the achievements of the nations of the world are measured and reported on in the HDI, the Scandinavian countries consistently rank very highly. Just how highly, we will see later. High rankings on the HDI and in other global measures suggest a major transformation of the Scandinavian countries since the Viking period of 1000 years ago.[2] The Scandinavian peoples began their journey as some-time rapists, pillagers and pagan seafaring explorers, yet they now embody Enlightenment principles and are often described as humanist, free, democratic, honest, caring, charitable, rationalist and liberal. We should not overlook the other Nordic country, Finland, as its society has similar characteristics. There has to be a story in this transition, and one that helps us understand a way forward towards sustainable development.

The task of unraveling over 1000 years of history is beyond the scope of this book, so the best I can do is provide a potted history and identify ideas and events that help us to understand why the experience of these countries points to a positive future. We will focus on Norway, which, we have already seen, has led the HDI rankings for the best part of the first decade of the twenty-first century.

Norway's story, according to authors Rodney Tiffen and Ross Gittins, demonstrates the application of a fundamental principle of sustainable development: the Hartwick rule. It is that (at least) part of the profits (or rents) earned from the exploration of a non-renewable resource, such as fossil fuels, is to be invested in replacement earning assets, such as renewable energy sources. It is not necessary that the investment is in a direct substitute for fossil fuels, although given their certain depletion this would be wise.

Norway and the other Scandinavian countries have long been at the forefront of global environmental policy. We explore in this chapter how and why these countries in particular have achieved their much-envied status. What can we learn from their politics, economics and international initiatives that will help the rest of us move more decisively to a sustainable future?

A SHORT HISTORY

We know that about 14 000 years ago parts of the Norwegian coastline emerged from the Ice Age, during which the country had been uninhabitable. We know little of the people who settled the newly emerged land in 1200 BC. Øivind Stenersen and Ivar Libaek point out that 'We shall never know if the first people to cross the Norwegian Channel used boats or migrated across one cold winter's day, but there is much to suggest that they were tempted northwards by bounteous food resources'.[3]

Move forward 2000 years and we know a little more. Gwyn Jones, whose *History of the Vikings* is one of the best of numerous books on the subject, writes that we can go back 12 000 years to the earliest post-glacial period in Scandinavia, when 'men were moving over its habitable areas, food-gathering, hunting, fowling, and fishing, leaving their mark on a flint here, an antler there, in Denmark by Bromme north-west of Soro in Zealand, in Sweden in Skane and Holland, in Norway in Otsfold … and … in the south-western coastal region, and along the west coast from Bergen to Trondheim'.[4] The Norwegians were known as the Fosna folk. Jones continues:

> *It is meaningless to talk of nationality in those distant times, and idle to speak of race; but these hunters, fishermen, and food-gatherers … who knew, or over the centuries came to know, the bow and arrow, knife, scraper, harpoon, and spear, who developed the skin-boat, would possess the first known tamed animals, the big wolf-like dogs of Maglemose and Svaerdborg, and buried their dead in the shallow graves in close proximity to the living – these were the parent Scandinavians.*[5]

Stenersen and Libaek remind us, if we need reminding, that skis and sleds were part and parcel of everyday life, allowing migrations over hundreds of kilometres over a period of a few months.[6]

Farming developed in Scandinavia around 4000 BC; however, it is believed that hunting and fishing remained the more important activities for a considerable time after farming began. By 3000 BC there had been a widespread occupation of Norway. This is when the Battle-Axe people arrived, bringing with them, it is thought, the Indo-European language and new 'technology': no longer would axes be made by endlessly chipping away at the stone; now the stone was ground. This new tool gave the nomadic hunters and gatherers the means to fell forests (of which Norway had and still has an abundant store) and become farmers. The next major influence came in about 500 BC with the introduction of the Iron Age to Norway. At the time, 'the whole country ... became populated as this rather short, long-headed Nordic type mingled with the taller survivors of the Stone Age people to produce what is still the typical Norwegians'.[7]

If we rely on the historical records, rather than the findings of archeologists, we have to wait until circa 300 BC for Scandinavia to get a mention. Greek geographer Pytheas of Massalia is believed to have sailed for six days north of Britain to a land close to the Arctic Circle that was, in his view, inhabited by barbarian agriculturalists. As early as this, the Scandinavians were in contact with the (then) advanced societies of the Mediterranean. The Norse and the Romans fought battles (which the Norse won) in Gaul, Spain and northern Italy. The Norse moved south, the Romans north, cultures made contact and resources were sought and obtained – by trade or the sword. The mixing of cultures that underpins European society from Greece (if not further east) to Scandinavia had begun.

The great migrations

By the time of Nero's reign (AD 54–68), reliable reports from Mediterranean explorers who travelled to the north identified a large 'island', called Scandza by them. From this we get the modern term 'Scandinavia'. Even then, 700 years before the Viking era, it was reported that the Norse had ships with a prow at each end. Also well before the Viking era was significant movement of Norse (and other peoples) throughout Europe. This is the era of the 'great migrations', as Jones describes.

> *The Visigoths, Eruli, Ostrogoths, and Langobards forced their way into Italy; the Franks, Visigoths, Alamanns, Bavarians, and Burgudians partitioned and repartitioned Gaul; from Gaul the Visigoths swung south to conquer Spain, and the Vandals moved on by way Andalusia to North Africa; in the mid-fifth century Angles and Saxons, some Jutes and Frisians, left their homes in and near the Danish part of Scandinavia to transform Roman Britain into Germanic England.[8]*

These migrations were extremely important to the mixing of cultures. Something very similar happened in ancient Greece and Persia. Numerous theories attempt to explain the rise and rise of Europe – notwithstanding centuries of bloody and intensive warfare – but there is a story in the free mixing of people and ideas, which closed societies cannot achieve.

What the Scandinavian contribution was in these migrations is not clear, although the pattern of conquest is consistent with how the Scandinavians in the Viking age proper (three to four centuries later) behaved – the colonial ventures, the campaigns and the raids were similar. The Scandinavian raiders and explorers not only returned home with precious stones, metals and slaves, but also ideas – which changed their societies.

Cultural and ethical mixes

Who is a Spaniard, an Italian or a Brit (to name three of the dozens of twenty-first century nationalities) if not a descendant from the melting pot of the great migrations or from the migrations that had occurred before or those that were yet to come, including the Viking era raids and settlements? How is that we fight on nationalistic or ethnic grounds when we all come from the same pot – as does a good stew or curry? Was Karl Marx right in asserting that the real fight was over resources and anything else was contrived as an excuse to go into battle?

Not to overstate the importance of the southward raids and migration by Norse peoples, we must recall that, for four centuries after the birth of Jesus, the Roman Empire was 'the economic and political powerhouse of Europe',[9] bringing the Latin alphabet and Roman measurements. The Scandinavians obviously learnt from the Romans. Trade, from north to south and south to north, was substantial and, what to the Scandinavians were luxury goods (gold and silver jewellery, bronze vessels and glassware), headed north and furs, animal skins and slaves (presumably captured during Scandinavian raids in other parts of the globe) headed south.

The raiders come

'It is some 350 years that we and our forefathers have inhabited this lovely land, and never before in Britain has such terror appeared as this we have now suffered at the hands of the heathen.'[10] This is how Alcuin, an eighth-century scholar in Charlemagne's court, described the Viking raid on the monastery at Lindisfarne in 793.

Written in the *Anglo-Saxon Chronicle* that same year was the entry:

> *In this year terrible portents appeared over Northumbria and sadly affrighted the inhabitants: these were exceptional flashes of lightning, and fiery dragons were seen flying in the air. A great famine followed soon upon these signs, and a little after that in the same year on the ides of January [June] the harrying of the heathen miserably destroyed God's church in Lindisfarne by rapine and slaughter.*[11]

By the beginning of the ninth century, Western Europe had entered a dynamic period of its history. Trading and raiding were changing Europe from the Mediterranean to the Arctic Circle. For example, in AD 860, the Vikings attacked Constantinople – they had reached there by AD 839 – and within decades it had become a major market for the Viking Rus.[12] The end of the western Roman Empire opened the door for those willing and able to fill the vacuum it left. By the time it became possible for a map-maker to make some sense of the linguistic and ethnic groupings in northern Europe, the Viking era had well and truly commenced.

> *The relative stability of western Europe under Charles the Great (Charlemagne) and of England under Offa of Mercia was shattered in the ninth century by attacks by Saracens in the south, Magyars in the east and Norwegians and Danes in the north and west. The Saracens pillaged Rome in 846 and … in 890 raided deep into southern Gaul. Northern Italy and Germany were prey to the Magyars … The Vikings of Norway and Denmark also began as raiders; but in their case an initial phase was followed by settlement and colonization, first in Orkey and Shetland, then in Ireland where Dublin was founded c.841, later (c.870) in Iceland and in England, where the Danish armies occupied the countryside … the Midlands after 876. In France the West Frankish king conferred the lands at the mouth of the Seine – the later duchy of Normandy – on the Danish leader Rollo in 911.*[13]

This quote summarises the start of hundreds of years of fascinating history that show both the brilliance and brutality of the Vikings.

Viking rulers

Harold Fairhair rules Norway

Norway as a nation first appears in the Viking age. In about AD 900, Harold Fairhair imposed his rule over all of the country. His kingdom had a short life: 'The turbulent Viking chiefs, long accustomed to assert their will by the sword at home as well as abroad, soon regained their local independence.'[14] Domestic politics aside, there was a period when the kingdom of Norway included not only the country we know today, but also large parts of central and western Sweden, much of Lapland in neighbouring countries, the Faroe Islands, Iceland, Greenland, Vinland (the first European settlement in North America), parts of Ireland, the Isle of Man, the Hebrides, northern Scotland, the Orkneys and the Shortlands. This was no small 'empire' – Norway was twice its present size.

While we might marvel at the courage and seafaring skills of the people who achieved this spread of possessions, it is worth keeping in mind that 'the ordinary Norwegian farmed and fished and hoped his quarrelsome betters [the Viking chiefs] would leave him in peace. The happiness as well as the unity of Norway suffered from [the chiefs'] brilliant, brutal forays'.[15]

Olav the Fat

Olav the Fat (Olav Haroldson), a Viking born in Russia, was the first king after Harold Fairhair to bring Norway under one ruler – brief though this was to be. Olav followed a succession of great and famous characters with names suited to their personalities and physical characteristics; for example, Hakon the Good and Harold Bluetooth.

Olav founded Nidaros in AD 997. It is now Norway's third largest city, Trondheim. Olav attempted to get the pagan Vikings to convert to Christianity, but the local chiefs were not interested and sought to have Olav replaced as king. The Trondheim chiefs called on Canute the Great of Denmark and England[16] for help. He was a pagan with some claim to the throne of Norway. Olav was deposed, but not happily, and as expected sought to regain his crown, an unsuccessful venture as he was killed in AD 1030. While this would be the end of the story for mere mortals, the deceased Olav became, within a year of his death, the national saint of Norway and a symbol of national unity. In AD 1038, a treaty between the king of Norway, Magnus, and the King of Denmark, Hardeknut, acknowledged the two separate countries – with an agreement that whoever outlived the other should take over the other's kingdom. This was the way things were done then. Kingdoms were lost or gained in the roll of a dice. Norway was one day a nation in its own right, the next part of a union of two or more nations (if we include England), only to return to its original state some distant time in the future.

As was often the case in the Viking world, it wasn't long before troubles brewed and swords were drawn. Arthur Spencer explains what happened when the Norwegian–Danish agreement had to be enforced.

> *The quarrels this agreement caused on the death of the Danish king brought onto the scene, first in support of the Danish faction, then as joint King of Norway, the great warrior Harold Hardrade. He is famous in English history books for his typically Viking exchange of defiance with his cousin, Harold of England, about the amount of English land he was entitled to before the Battle of Stamford Bridge where, in 1066, he met his death. In Norway he is remembered for the foundation of Oslo [which was in time to replace Nidaros, or Trondheim, as the capital]. His two sons, Magnus and Olav the Peaceful, succeeded him and ushered in a period of prosperity and comparative calm.*[17]

Christianity comes to Norway

The last of the line of Harold Fairhair, Sigurd I, died in AD 1130 and a long period of disorder, intrigue and, at times, armed rebellion followed. The domestic troubles provided an opening for a Christian crusader to peddle influence. In 1154, an English cardinal by the name of Nicholas Breakspear (not an inappropriate name for a Christian peacemaker) established the Archbishopric of Trondheim in central Norway. When the new king, Magnus V, was crowned in 1163, the archbishop performed the ceremony – Norway's ruler, if not its people, had become Christian. Then in 1184, a pretender to the throne, Sverre from the Faroes, overthrew Magnus. Sverre set about strengthening the rule and growth of a centralised government. However, central government threatened the power and privileges of the newly installed church, so the clergy did not turn the other cheek or give in to Caesar but rather fostered armed rebellion against Sverre and his successor. Calm had returned to the country by 1240, under the leadership of King Hakon IV. A period of prosperity set in. Hakon's son, Magnus the Law-mender, continued the good work of his father.

Significant events

The Norwegian empire was consolidated, trade grew (particularly with Britain), towns expanded, and a period of artistic endeavour ensued. Then at the end of the thirteenth century a significant economic event happened: the north German Hansa merchants established their power in Scandinavia and so Norway's traditional trading relationship with Britain suffered. By the middle of the next century, the Black Death had reached Norway, killing about two-thirds of the country's people. To make matters worse, at about the same time, the climate deteriorated in one of the world's normal climatic perturbations. It was not a happy period.

The event that had the most significant political impact on Norway came with the end of the direct male line of the royal family in 1319. The son of a Norwegian princess inherited, at the age of two, the Norwegian throne. He was also the heir to the Swedish throne. From that date onwards – until 1905 – Norway was one part of the various unions formed between the Scandinavian countries. In 1397, the Union of Kalmar established the three separate kingdoms of Norway, Denmark and Sweden, united under one sovereign. The Swedes broke from the union in 1523, but Norway and Denmark remained united until the Peace of Kiel in 1814. The union was by no means one of equals. Norway was effectively a Danish colony; in fact, Denmark went as far as declaring Norway a Danish province in 1536. One consequence of this was that Norway become involved in Denmark's wars and as a result lost a considerable part of its eastern lands to Sweden.

The Napoleonic War

In the Napoleonic War, Denmark made the mistake of siding with France. Norway wanted out of the union and looked to a new partnership with Sweden. Jean-Baptiste Bernadotte, formerly a French general who had served under Napoleon, took over the Swedish throne in a peaceful revolution, then attacked and defeated Denmark. Bernadotte had a bit of trouble in reconciling his view of the world with that of the egalitarian Swedes. Box 1 explains why.

> **Box 1:**
> **Culture shock**
>
> When the new king [Bernadotte] was installed, he addressed the Swedish parliament in their language. His broken Swedish amused the Swedes, and they roared with laughter. The Frenchman who had become king was so upset that he never tried to speak Swedish again. Bernadotte was a victim of culture shock: never in his French upbringing and military career had he experienced subordinates who laughed at the mistakes of their superior. Historians tell us he had no more problems adapting to egalitarian Swedish and Norwegian mentality (he became king of Norway as well) .... He was a good learner (except for language), however, and he ruled the country as a highly respected constitutional monarch until 1844.[18]

A new constitution

In the Treaty of Kiel, Denmark lost control of Norway but kept the then Norwegian dependencies: Iceland, Greenland and the Faroes. Norway was forced into a relationship with Sweden in an arrangement in which it had no say. The Norwegians objected and set about re-establishing Norway as a completely independent country. A new Norwegian constitution was promulgated on 17 May 1814 – to become Norway's national day – but soon after, in July, Sweden invaded Norway. The hostilities were of short duration and a new formal union between Sweden and Norway was agreed to by the parliaments of both countries. In 1905, this union ended and Norway elected Hakon VII as king in a referendum.

Rich and social democratic

In more recent times, Norway has become a very rich social democratic country, with its economic base originally built on shipping, paper and fish, then expanding to aluminum production, ferro-alloys and now oil (Norway is the fourth largest net exporter in the world). A resource-rich country keen on sustainable development, it was the first country in the world to establish a ministry of the environment. That was in 1972 and was the start of a number of environmental 'firsts' for the country.

HOW SCANDINAVIA COMPARES TO OTHER COUNTRIES

Although we have concentrated on Norway, the story is not that different in the other Scandinavian (or Nordic) countries. Hence we can treat them as one in the following comparison to various other countries. They all rank highly on various criteria that are used to define good environmentally conscious and sustainable societies. What underpins this? What are the social, political and economic values and attitudes that lead to these countries outperforming the rest of the world? We are fortunate to have relevant and quite comprehensive recent data about the top eighteen industrialised countries in the world that compares social and economic patterns.

Political colours

Let us start with political ideologies. I have selected eight of the eighteen countries in the data set so to present a geographical, economic and cultural spread (see Table 1). Consider the political orientation of the national governments of these countries from 1950 to 2000. The three Scandinavian countries (Iceland is not in the list) are traditionally social democratic. Sweden is the most social democratic country in the group, followed closely by Norway. The USA, however, has no social democratic blood running through its veins and, in Australia[19] and the UK, there is a noticeable percentage of social democratic voters, but conservative government is twice as likely.

Table 1: Colour of governments, 1950–2000 (% of voters)[20]

RANK	COUNTRY	SOCIAL DEMOC-RATIC	CENTRE	LIBERAL	CONSER-VATIVE	OTHER
1	Sweden	77	10	7	4	3
2	Norway	72	12	4	12	0
4	Denmark	55	4	26	14	2
5	Australia	31	0	0	69	0
6	UK	30	0	0	68	2
7	Finland	30	33	12	10	15
12	Italy	22	63	7	0	8
17	USA	0	45	0	55	0

Income inequality

The next indicator is income inequality, as measured by the Gini coefficient, which indicates the divergence from complete income equality (0.1).The data in Table 2 shows that the four Nordic countries are the fairest in income distribution and have coefficients closer to 0.1 than the others, particularly the UK and the USA. A very similar pattern emerges if the ratio of rich to poor is considered – the ratio of incomes at the 90th percentile to those at 10th percentile. The four Nordic countries rank one to four, with Denmark moving from sixth place in Table 2 to fourth place.[21]

Table 2: Inequality[22]

RANK	COUNTRY	GINI COEFFICIENT
1	Sweden	0.221
2	Norway	0.238
3	Finland	0.247
6	Denmark	0.257
13	Australia	0.311
15	Italy	0.342
16	UK	0.345
17	US	0.368

Higher social capital

If we turn our attention to membership of trade unions, another feature of economic life, we notice a stark difference between the Nordic countries and the others (see Table 3). The decline in trade union membership over the past twenty-odd years has been dramatic in Australia (from 50 per cent to 25 per cent)[23] and the UK (from 52 per cent to 29 per cent), while it has remained constant or even increased (as is the case in Finland) in the Nordic countries. As all the Nordic countries are wealthy, and income inequality is low, there is no need for a strong trade union base to carry the fight for the working class. This suggests that the high membership must be a function of community spirit – in other words, high social capital.

Table 3: Membership of trade unions, 2000[24]

RANK	COUNTRY	EMPLOYEES (%)
1	Sweden	82
2	Denmark	76
3	Finland	76
4	Norway	76
13	Italy	35
15	UK	29
16	Australia	25
17	USA	13

Free education for all

Let us next consider commitment to education. While it does not necessarily follow that education has to be provided for by taxes and it can be a matter for the individual to fund personally, there is, on the face of it, a correlation between public expenditure on education and a society's commitment to educating its people. Voters support taxes going to education (hopefully, a good-quality education system) because they believe this is the most cost-effective way of building social capital – in other words, a good society. And, if there is no commitment to compulsory education (for a certain number of years), some children will miss out (through parents' neglect, ignorance or indifference). From Table 4, it is clear that all the selected countries are committed to public schooling.

For reasons that are not obvious, Australia stands out as the country least committed to funding its schools. Only the Nordic countries show a strong commitment to public funding of tertiary education. Australia and the USA are serious laggards, but for Australia this is a recent change. In the early 1970s, a social democratic (Labor) government abolished fees for university study; however, a more recent Labor government reintroduced fees and the conservative governments since then have increased the fees.

Table 4: Public share as a percentage of education expenditure, 2000[25]

RANK	COUNTRY	SCHOOLS	TERTIARY	TOTAL
1	Norway	99	96	99
2	Finland	100	97	98
3	Sweden	100	88	97
4	Denmark	98	98	96
10	Italy	98	78	91
12	UK	89	68	85
15	Australia	85	51	76
17	USA	90	34	68

Reading and writing

The conventional wisdom – as recognised in the HDI – is that literacy is a key indicator of a mature society. One aspect of literacy is the ability to be informed about and be interested in world, national and local events. Undoubtedly newspapers provide the most comprehensive and analytical coverage of the news, although television, radio and the internet can also provide good news coverage (although they don't always).

Table 5 shows data on newspapers sold per 1000 people (or per capita, if you undertake the necessary arithmetic). Norway is well ahead of the field – while there have been fluctuations over the 30-year period from 1970 to 2000, including a decline from 1990 to 2000, overall there has been a considerable increase in newspaper circulation. The only other country to show an increase – in its case, a very small one – is Finland. On the other side, there has been a dramatic fall in newspaper readership in

Australia. Newspaper readership is only half of what it was 20 or so years ago, with the greatest fall occurring after 1990. To fully comprehend this it would be necessary to analyse the demographic trends in readership, but if the story is that older readers are not being replaced by younger readers, this is of serious concern. Possible explanations – such as greater use of the internet in Australia than in other countries – do not stand up to scrutiny. The top internet users are the Scandinavian countries – and they continue to read newspapers.

Table 5: Daily newspapers sold per 1000 population[26]

RANK	COUNTRY	1970	1980	1990	2000
2	Norway	397	463	610	573
3	Sweden	539	528	526	464
4	Finland	434	505	558	443
6	UK	453	417	388	320
9	Denmark	363	366	352	279
12	USA	296	270	245	197
14	Australia	321	323	305	162
18	Italy	144	101	105	110

Turn off the TV

It might be expected that where societies don't have a high newspaper readership they are likely to have a higher propensity for television viewing – and so it is. The populations of the four Nordic countries are watching the least amount of television. Yet it seems that Australians do not fully compensate for their lack of interest in newspapers by watching a substantial amount of television. This could be explained by the fact that Australia's climate permits, actually encourages, an outdoor lifestyle. Neither television nor newspapers are substitutes for the 'barbie' or the football! Also, to confound understanding of general literacy standards in Australia, the country has a high book readership.[27]

Table 6: Average weekly household television viewing, 1990s[28]

COUNTRY	AVERAGE HOURS PER WEEK
UK	28
USA	28
Italy	27
Australia	22
Denmark	20
Finland	18
Norway	18
Sweden	18

Sustainability measures

We now come to a number of variables that, both individually and in combination, give us a reasonable idea of the strength of social capital and, indirectly, commitment to sustainability in our chosen countries. Table 7 shows the results for life satisfaction measures (rated on a scale of 1 = not at all satisfied to 10 = very satisfied). Of the Scandinavian countries, Denmark rates the highest (as well as rating high overall). Sweden and Finland do less well – although still better than Australia, the UK and the USA.

Table 7: National surveys on life satisfaction[29]

COUNTRY	YEAR OF SURVEY	LIFE SATISFACTION
Denmark	1999	8.24
Australia	2005	7.28
Sweden	2006	7.74
UK	2006	7.60
Norway	2007	7.96
Finland	2005	7.84
USA	2006	7.57
Italy	2005	6.89

Trust in others

One of the fundamental variables underpinning social capital is trust in others. Norway has the most trustworthy population, followed by the other Nordic countries. Falling below the 50 per cent line are all the others (USA, Australia, UK and Italy), as shown in Table 8. These are dismal results in countries whose people take their ethics (so we thought) from the progressive Enlightenment thinkers.

Table 8: Responses about whether others can be trusted, 1990s surveys[30]

RANK	COUNTRY	AGREE PEOPLE CAN BE TRUSTED (%)
1	Norway	65
2	Sweden	63
3	Denmark	58
5	Finland	54
8	USA	44
10	Australia	40
12	UK	38
14	Italy	35

Pride in what?

When we come to consider national pride we see data that, at first glance, is counterintuitive. Citizens of the USA are the most proud, even though the country ranks very low on equality, membership of trade unions, public spending on education, confidence in its legal system and in its parliament. On the other hand, the four Nordic countries (and Italy) all fall below the 50 per cent mark; that is, less than half the population in these countries indicated that they were very proud of their country. This raises a very interesting question about the nature of being proud of your country – it can result from something well deserved, such as winning a major international sporting event, or being powerful and rich, a position that might or might not be associated with hard work rather than good fortune (endowed natural resources). It can also be associated with being parochial or even arrogant. If a country's people are not very proud this might suggest that they are more reserved or more reflective about their achievements or more outward-looking, giving credit where credit is due. Significant attitudinal differences come into play.

Putting your life on the line

A fascinating exercise is to compare pride (Table 9) in your country to willingness to fight for your country (Table 10). The Nordic countries jump back to the top when willingness to fight for your country is the issue. One possible explanation is that their populations are proud and loyal while not being overly proud or being proud for the wrong reasons.

Table 9: Responses about pride in nationality in 1990s surveys[31]

RANK	COUNTRY	VERY PROUD (%)
2	USA	77
3	Australia	70
5	UK	53
6	Norway	48
7	Finland	44
8	Sweden	43
9	Denmark	43
10	Italy	40

Table 10: Responses about willingness to fight for your country, 1990s surveys[32]

RANK	COUNTRY	WILLING (%)
1	Norway	90
1	Sweden	90
3	Denmark	89
4	Finland	86
5	USA	78
6	Australia	75
7	UK	74
17	Italy	31

Confidence in social institutions

Tables 11 to 14 relate to the confidence the citizens of the respective countries have in key social institutions. The USA comes out number one in having confidence in the church – more so than Italy – but the other countries seem to pride themselves on their secularity.

Confidence in private companies provides a mixed result and it is difficult to explain the results without more data. Confidence in the legal system and parliament finds the Nordic countries ranking highly. Surprisingly, Australians do not seem to have a great faith in the legal system or their parliament. Confidence in the civil service does not follow the patterns we have noted above and it would need more detailed probing if we were to make sense of these results.

Table 11: Confidence in the church[33]

RANK	COUNTRY
1	USA
1	Ireland
3	Canada
4	Italy
5	Belgium
6	Austria
7	France
7	Norway
9	Denmark
9	Finland
11	UK
12	Australia
12	Sweden
14	Switzerland
15	Germany
16	Netherlands
17	Japan

Table 12: Confidence in companies[34]

RANK	COUNTRY
1	France
2	Italy
3	Australia
4	Norway
5	Ireland
5	USA
7	Canada
8	Belgium
9	Netherlands
10	Finland
10	UK
12	Switzerland
13	Austria
14	Sweden
15	Denmark
15	Japan
17	Germany

Table 13: Confidence in the legal system[35]

RANK	COUNTRY
1	Denmark
2	Norway
3	Finland
3	Switzerland
5	Netherlands
6	Japan
7	Sweden
8	Austria
8	France
10	Germany
11	Canada
12	UK
13	USA
14	Ireland
15	Belgium
16	Australia
17	Italy

Table 14: Confidence in government and civil service[36]

CONFIDENCE IN PARLIAMENT		CONFIDENCE IN CIVIL SERVICE	
RANK	COUNTRY	RANK	COUNTRY
1	Norway	1	Ireland
2	Netherlands	2	USA
3	Ireland	3	Denmark
4	France	4	Canada
5	Switzerland	4	Switzerland
6	Sweden	6	France
7	UK	7	Norway
8	Denmark	8	UK
8	Belgium	9	Netherlands
10	Austria	10	Sweden
11	Germany	11	Belgium
12	USA	11	Austria
13	Canada	13	Australia
14	Finland	14	Germany
15	Italy	15	Japan
15	Australia	16	Finland
17	Japan	17	Italy

THE ENVIRONMENT

One issue that interests us even more than those we have been discussing is attitudes to the environment. The three Scandinavian countries top the list in Table 15, but Finland features poorly. Australia and the UK were not too far behind the Scandinavian countries. However, note that the 2005 Environmental Sustainability Index ranked Finland, Norway, Uruguay, Sweden and Iceland one to five. The ranking scheme used in Table 15 and that of the Environmental Sustainability Index deal with considerably different attributes of environmental care and this explains the differences. However, the overlap of the Scandinavian countries suggests a strong commitment to a healthy environment. Finland remains an enigma.

Table 15: Willingness to pay more taxes to prevent environmental damage, 1990s surveys[37]

RANK	COUNTRY	WILLINGNESS TO PAY (%)
1	Sweden	80
2	Norway	74
3	Denmark	70
4	Australia	69
5	United Kingdom	68
8	United States	60
11	Finland	55
11	Italy	55

THE RISE AND RISE OF SCANDINAVIA

How to summarise the rise and rise of the Scandinavian countries? The role that class plays in determining social relationships might be the clue. Norway in particular – but also the other Scandinavian countries – has experienced generations of 'classlessness'. This idea – it is as much an ethical idea as a fact of economic life – is strongly supported by its people, regardless of which political party they support. A commitment to education, the exploration of ideas (as much as exploration of other places in the world) and independent endeavour are also fundamental cultural attitudes of the Norwegians.

Why do a particular people set the pace at a particular time in history? Why was there the great Greek era of philosophy commencing 500 years before Christianity? The thing we know about Greece with certainty is that a surplus of available material goods allowed a scholarly class to exist. Why did the Enlightenment of the eighteenth and nineteenth centuries in England and Scotland take place? Again, what we know with certainty is that the economic system of that time produced a surplus sufficient to allow for the existence of a scholarly class.

Today the industrialised world produces an unbelievable surplus but scholars are very poorly paid compared to financial speculators and the so-called managerial class. Education tends to be directed to instilling workplace skills at the expense of scholarship and exploring big ideas. Advertising and the artificial glamour of celebrity life works against the concept of seeking a happy, contented life. It was the Norwegian-American economist Thorstein Veblen who mocked the vacuous lifestyles of those forever attempting to be one step ahead of the 'Jones' via conspicuous consumption – a no-win game. The Norwegians, and their fellow Scandinavians, are not perfect – no-one is – however, at this point in time, they are writing the script for a good sustainable future.

The Nordic enlightenment

Viking age Norwegian farmers originally owned their land. Over time, the bulk of farmers became tenant farmers. The king, the church and a chiefly class came to own about 70 per cent of the land.[38] This situation resulted from the king taking land from those who opposed him and lost – the usual story worldwide in the pre-democratic era. However, 'Norwegian farmers enjoyed greater freedoms than farmers who lived on large manors in other parts of Europe … The Norwegian farmer was legally free throughout the whole of the Middle Ages and he was not seen as socially degraded because he was a tenant; a tenant farmer on good land could be in a better position than one who owned his own farm'.[39]

In 1821, well before the European revolutions of 1848, the Norwegian parliament (the Storting) passed legislation – against the king's will – to abolish all noble titles and privileges. In fact by 1833, the Storting comprised a majority of farmers and was called the 'peasant parliament'.[40] In Sweden, peasant farmers had their own 'estate' in parliament since the fifteenth century.

The Scandinavian countries, more so Norway than the others, were remarkably different from the rest of the world. Oystein Sorensen and Bo Strath write in some detail about the Nordic enlightenment and make this critical point – it 'provoked not a revolution but just enlightenment [and] ironically and paradoxically enough had the peasant as its foremost symbol'.[41] A peasant as the source of free-thinking endeavour and the person who was best suited to develop a progressive society – really![42]

Nordic society

History matters. Norway, in particular, but the Nordic countries generally, developed into modern nations from a democratic and classless base. Sorensen and Strath describe the historical key elements of the Nordic model of society: 'the peasants, education … the nation (folk) as a community of destiny, a social liberalism conducive to forming coalitions with the social democrats, profitable capitalism, and the political production of welfare. The core of the idea is the combination of equality and individual freedom'.[43]

Progressive politics

The twentieth century saw the development of modern Norway, the key characteristics of which became evident early on. An extremely strong commitment to peace was paramount. Norway and other Scandinavian countries adhered to a policy of neutrality in World War I, although their sympathies were with the Allies. After the war, Norway, as with the rest of the industrialised world, suffered the Great Depression. Very early on, a Keynesian approach to economic management gained popular support. The Labour Party under Johan Nygaardsvold was voted into office in 1935 with the goal of bringing the economy under state control. This government remained in office for ten years, the last five of which were spent in exile in England during World War II when Germany occupied Norway.

The German invasion

When World War II commenced in September 1939, Norway declared its neutrality, consistent with its belief in the League of Nations and the use of peaceful methods to resolve conflicts. Nazi Germany was in no mood to recognise such niceties – on 9 April 1940, without declaration of war, Nazi Germany invaded Norway. The small Norwegian armed forces were no opposition to the far greater German forces. The government and the king went into exile in England, while the Norwegian resistance organised at home.

Norwegians resolutely and virtually unanimously rejected the pretentious Nazi Germany's claim to Nordic brotherhood. The resistance was the evidence the Germans needed to underscore their rejection. The resistance was to have some spectacular results, based on pure courage and skill. As Spencer writes, 'After many years of peace the Norwegians regained the warrior spirit of their terrible Viking forebears'.[44] The Germans were to discover this to their dismay. While the resistance kept the Germans occupied at home, the very large Norwegian merchant navy, which fortunately was at sea or abroad when Norway was invaded, made a significant contribution to the Allied cause. Norway had the third largest fleet in the world before the war, and its merchant fleet was the most modern. It lost half its ships and thousands of sailors during the war with Germany.

To the left

The war ended with Nazi Germany surrendering on 7 May 1945. Politically, the Norwegians moved further to the left after the war and Labour governments retained power for long periods. Norway was as committed to the idea of the United Nations as it had been to the League of Nations. Therefore, at the close of World War II, it became a founding member of the UN and provided the first secretary general, Trygve Lie.

Norway's support for the principles of this new body went far beyond attending the mandatory meetings. It always paid its dues and has consistently paid more than most countries. Today, only Denmark and Norway meet the criteria, adopted in principle but not in practice by all donor countries, of providing .07 per cent of GDP as foreign aid. Norway has, from the word go, been willing to provide personnel for peacekeeping. In fact, during the Cold War, Norway saw itself as playing the essential role of building a bridge between the two power blocks, the West and the Soviet Union. However, events in Europe, such as the communist coup in Czechoslovakia and Soviet pressure on Finland, undermined Norway's attempts. Reluctantly Norway joined NATO in 1949.

Right and left see eye-to-eye

Norwegian party politics are not readily comprehended by people used to a distinct left–right dichotomy or strong class divisions. In describing the Norwegian Liberal Party, as it was prior to World War II, Stenersen and Libaek write, 'The party wanted to remain above class differences and use the state to help the weak. The Liberal Party believed in social reforms, price controls'.[45] The similarity between the left and right Norwegian political parties is illustrated by the fact that in the 1945 parliamentary elections all parties put the same policies to the electorate. The war experience and the resistance consolidated the existing classlessness of Norwegian society.

Marching to a different drum

The other Scandinavian countries had somewhat similar but nevertheless specific experiences in the period over which they became social democratic states. Finland, as a Nordic but not Scandinavian country, was caught up throughout much of the twentieth century in its difficult relationships with both Russia and Germany. Iceland, somewhat removed by vast stretches of the Atlantic Ocean from the rest of Europe, probably had the easiest journey to modernity – notwithstanding three so-called 'cod wars' with the UK. Sweden, neutral throughout both the world wars, experienced the letting of some blood when, at the Adalen labour riots in 1931, troops killed five citizens. From that day on, for 40 years without a break, the Social Democratic Party ruled in Sweden.

Without doubt the Scandinavian countries are different to the rest of the world. While relatively small in area and population, they have an elevated role in world affairs. In boxing parlance, they punch above their weight. Why? History helps, as the example of Norway shows. Although Norway did have its powerful Viking chiefs, it never had an aristocracy in the way that England and other parts of Europe had. More significantly, it never had a powerful bourgeoisie, as developed after the Industrial Revolution in many parts of the industrialised world. Furthermore, a mandarin class did not arise at any time in its history.[46] It was and still is the closest thing there is to a classless society – yet it is capitalist.

1   Rodney Tiffen and Ross Gittins, *How Australia Compares*, Cambridge University Press: Port Melbourne, 2004, p. 93.

2   The convention is to date the Viking era from AD 793 to 1066. Gwyn Jones (*A History of the Vikings*, 2nd edn, Oxford University Press: Oxford, 1984) writes of the 'Viking Age' proper being circa 780 to 1070.

3   Øivind Stenersen and Ivar Libaek, *The History of Norway from the Ice Age to Today*, Dinamo Forlag: Norway, 2003, p. 7.

4   Jones 1984, p. 17.

5   Jones 1984, p. 17, notes that this ancient hunting culture (or parallels to it) survive today among not only the Lapps (the Sami) of Finmark but also among the Norse people of Norway. She also tells us that the three Scandinavian countries – Norway, Sweden and Denmark – have been known for more than 1000 years by these names.

6   Today, we are only likely to see skis at tourist resorts and in the Winter Olympics, as they are no longer used as transport in most of Scandinavia. It was a different story 50 years ago; I vaguely recall skiing to school in northern Norway.

7   Arthur Spencer, *The Norwegians: How they Live and Work*, David and Charles: United Kingdom, 1974, p. 19.

8   Jones 1984, p. 28.

9   Stenersen and Libaek 2003, p. 15.

10   Quoted in Jones 1984, p. 194.

11   Quoted in Jones 1984, p. 195.

12   'Bus' refers to the Vikings who ruled Russia (Sawyer, *The Oxford Illustrated History of the Vikings*, Oxford University Press: New York, 1997, p. 14).

13   *Collins Atlas of World History*, Collins: United Kingdom, 2004, p. 50.

14   Spencer 1974, p. 19.

15   Spencer 1974, p. 20.

16   Canute was king of Denmark, England and Norway at one stage.

17   Spencer 1974, p. 201.

18   Geert Hofstede and Gert Jan Hofstede, *Cultures and Organizations, Software of the Mind: Intercultural Cooperation and its Importance for Survival* (2nd edn), McGraw-Hill: New York, 2005, pp. 39–40.

19   The social democratic party in Australia was in its early years called the Australian Labour Party, but in 1972 changed its name to the Australian Labor Party, reflecting the influence of the American labour movement. Cynics argue that the American spelling represents an acceptance of an American (laissez-faire) ideology.

20   Tiffen and Gittins 2004, p. 36. Note that rounding errors occur.

21   On the topic of equality, it is an interesting observation, if nothing else, that in Sweden and Norway people of different social groups are of the same height – which could say something about nutrition and health – but in the USA height differences are quite evident across social classes (Robert Fogel, *The Escape from Hunger and Premature Death, 1700–2100: Europe, America and the Third World*, Cambridge University Press: New York, 2004, p. 40).

22 Tiffen and Gittins 2004, p. 136.

23 However, despite the introduction of new industrial (anti-union) laws in 2006, union membership is on the increase.

24 Tiffen and Gittins 2004, p. 84.

25 Tiffen and Gittins 2004, p. 120.

26 Tiffen and Gittins 2004, p. 204.

27 It is not unknown for spectators to take a novel to an Australian Rules football match to read at half-time or when the play becomes boring.

28 Tiffen and Gittins 2004, p. 182.

29 The data comes from the 2005–2007 World Values Survey and the European Values Study, quoted in Ronald Inglehart, Roberto Foa, Christopher Peterson and Christian Welzel, Development, freedom, and rising happiness: a global perspective (1981–2007), *Perspectives on Psychological Science*, 3(4), Appendix A, pp. 282–4.

30 Tiffen and Gittins 2004, p. 204.

31 Tiffen and Gittins 2004, p. 204.

32 Tiffen and Gittins 2004, p. 204.

33 Tiffen and Gittins 2004, p. 204.

34 Tiffen and Gittins 2004, p. 204.

35 Tiffen and Gittins 2004, p. 204.

36 Tiffen and Gittins 2004, p. 204.

37 Tiffen and Gittins 2004, p. 204.

38 Stenersen and Libaek 2003, p. 37.

39 Stenersen and Libaek 2003, p. 37.

40 Stenersen and Libaek 2003, p. 81.

41 Oystein Sorensen and Bo Strath, *Cultural Construction of Norden*, Scandinavian University Press: Oslo, 1997, p. 1.

42 Mao Zedong thought he could build a 'communist' revolution and economy with the Chinese peasants. He failed.

43 Sorensen and Strath 1997, p. 7.

44 Spencer 1974, p. 27.

45 Stenersen and Libaek 2003, p. 107.

46 Sorensen and Strath 1997, p. 6.

10

AUSTRALIA: AUSSIE, AUSSIE, AUSSIE!

I should be so lucky.

Kylie Minogue

We have travelled the world and have learnt much. Yet we have neglected a fascinating environmental story – it is here in Australia. Australia is a country that punches above its weight in various international arenas. Sport comes to mind readily and wool should, as the country is the world's major producer and exporter of superfine Merino wool. Until the minerals boom, driven in recent years by demand from a rapidly developing China, Australia was known for 'riding on the sheep's back', so important was its export of wool. However, competition from synthetic textiles and cotton reduced the economic importance of wool. Fortunately, this happened just as international tourism to Australia started to increase and generate considerable foreign exchange and a vast number of jobs (not necessarily full-time or highly paid) in the hospitality sector. The importance of coal, iron and other mineral exports increased as well to the extent that mineral (including natural gas) exports dominate this sector.

WORLD RANK NUMBER 3

The minerals will eventually run out – but for coal this will be a long time in the future as there is so much of it. The threat of severe climatic events from a carbon-dioxide-induced greenhouse effect could bring the mineral boom to an end much sooner than physical depletion. Will Australia be able to retain its high rank as number three on the human development index (HDI) when this occurs, particularly if another industry is not phased in to replace mining? Are the significant rents (royalties) and profits from mining being invested in replacement industries rather than being spent on consumer goods and housing? If rents and profits are being invested, is the target sustainable industries that will provide incomes equivalent to the ones that run out of steam (no pun intended)? These are some of the many questions about Australia's environmental credentials that we need to ask, and attempt to answer.

SOME HISTORY

We are not certain when the first people came to Australia; however, it was approximately 40 000 to 60 000 years ago. They came via Indonesia and Papua New Guinea as part of a larger migration into the South Pacific region, initially from, the evidence suggests, southern China. They came a long time before the thaw that ended the last major ice age (about 11 000 to

15 000 years ago) and when movement into Australia was relatively easy. Then the seas rose.

We believe that there was not a great deal of migration into Australia or exchange between neighbouring peoples after the thaw. We know that there was some interchange of goods between peoples at the top end of Australia, with its close proximity to Indonesia and Papua New Guinea. At the other end of the country the people in Tasmania were completely cut off from the rest of the Aboriginal peoples – and the rest of humanity. Their complete isolation only came to an end when a few European seal hunters came in the late 1700s, some years before formal British settlement of Tasmania (at that stage still known as Van Diemen's Land) as a penal colony in 1803. It is difficult, perhaps impossible, for twenty-first century cosmopolitan people to put themselves in the place of the indigenous Tasmanians before the Europeans came. What would life have been like with no foreign raiders or traders, no source of new ideas for thousands of years? There were very few indigenous Tasmanians, about 5000-odd when the Europeans arrived, and Tasmania was their entire world. We know a considerable amount about many ancient societies through written records and archeological discoveries. This does not apply to the first Tasmanians – there is no record of the Aboriginal discovery and settlement of the island.

Colonisation

The official records show that the first Europeans to establish a colony in Australia arrived with the First Fleet, which landed in Botany Bay on 26 January 1788. Captain James Cook, who did much to alert the British government to the size and resources of Australia, had 'discovered' the country in 1770, but various other explorers, from different European maritime powers, had actually 'discovered' Australia much earlier. Such is the way histories come to be written when claims of ownership are at stake.

At the time of colonisation, there were few indigenous people (and there probably had been only a few throughout time) and they were scattered over the vast country. As a consequence of their isolation for millennia, the indigenous peoples had neither the technology nor the weapons that the Europeans had. (This story is somewhat similar to the North American and South American experience, where the superior weapons of the Europeans – and their introduced diseases – were to destroy the lives and the fighting spirit of the indigenous peoples.) Where Europeans and Aboriginal peoples in Australia came into conflict over resources there was bloodshed, which continued at irregular intervals in a number of places. From the day of the first landing, Australia had become a British colony under the firm control of the British naval commanders, yet they were to have as much trouble with bushrangers and free-willed and determined pardoned convicts and free settlers as they had with the indigenous Australians.

The population grows and grows

The most fundamental influence on the Australian natural environment has been the increase in numbers of Homo sapiens in a continent that previously had known only a small population. The growth of the population from 1788 to the present can be demonstrated in a few snapshots that show the varying pace of, first, migration to Australia and, second, natural birth rates. At federation in 1901 (when all the state colonies were united into the nation-state of Australia) the population was 3.8 million. At the commencement of World War II in 1939 the population was 7 million. By 1950, with the birth of the first (postwar) 'baby boomers', the population was 8.3 million. By 1970, when the baby boom had run its course, it was 12.6 million. In 2008, it is 21.3 million. This seems a small number given the vast size of the Australian continent, but keep in mind that most of it is semi-arid and arid – deserts cannot support many people.

A LITTLE GEOGRAPHY

Australia is the world's largest island continent. Most of the country is a desert, a fact few people, including this generation's coastal-clinging Australians, recognise. The existence of vast deserts is a real surprise for those people who know the country from the tourist promotions displayed in New York, London and Tokyo, which give the impression that the country is a huge rainforest, teeming with koalas and kangaroos, and that which is not rainforest is coral reef (an underwater kaleidoscope of either beautiful or mouth-watering fish, depending on your attitude to fish). Also highlighted in the tourism promotions is a big rock in the middle of nowhere known as Uluru (or Ayers Rock) and a high bridge framing a shell-like construction that, on entering, we discover is a large theatre, the Sydney Opera House. This is the Australia of glossy tourism posters – some of which is the real Australia.

We can't escape the reality of the deserts – and other similar facts. Today there are only patches of remnant rainforest remaining (much was removed to provide farming land). There is a significant stretch of snow-covered alps, an extremely long, narrow and fertile coastline (particularly on the east coast), a massive irrigated inland area in the south-east corner and swathes of pastoral land that gradually give way to gibber stone, which in turn become sand and occasional scratchy clumps of vegetation clinging with all their might to bare soil through near-constant drought. A massive country with a wide range of ecosystems, more than its share of unique animals (kangaroos, koalas, wombats and platypus, for starters) and a relatively small population – this is Australia.

Early environmentalism

For the European settlers of 200-odd years ago, Australia started as a series of penal settlements, prototypes of the cities that are now Sydney, Hobart, Launceston, Melbourne and Brisbane. Adelaide and Perth were different, being settled by free settlers. Before the advent of sanitation engineering, the disgusting state of creeks, ponds and rivulets caused much ill-health, premature death and loss of productivity among those attempting to build the infrastructure needed by a displaced group of Europeans. However, despite this hardship, through their hard work, the settlements continued to grow. Through convict labour, some high-quality buildings and infrastructure was developed and still exists today.

The convicts and free settlers brought their knowledge, lifestyles and habits to the new country. Some of these habits (translated into practices) were to prove disastrous in this southern hemisphere land of unique flora and fauna and mild to tropical climate – Australia really was a land down under in every possible sense. The early settlers also came with many vices – the rabbit, for example, was brought here so that 'aristocratic' Australian pastoralists could engage in the British 'sport' of hunting.

Because of the country's vastness and their tiny numbers, the new arrivals were not concerned about using watercourses to deposit all possible wastes or allowing the carcasses of dead animals to decay in the streets. The small numbers meant the waste was only noticeable in the crowded streets – yet concentrate even a few people in a confined area without the infrastructure for clean water and allow the disposal of their effluent in watercourses and the result is serious pollution. Eventually the downside of these polluting practices became all too obvious and ill-health, once thought a curse of the 'gods', could be traced to the pollution. Fortunately, enough literate and forceful citizens (mainly women) campaigned for change and, by the late nineteenth century, successful campaigns to clean up the towns and provide adequate sanitation and clean water bore fruit. These campaigners were the first environmentalists in Australia. And they were very successful, probably because they were dealing with very obvious cases of pollution and its impact on health.

Bringing Europe to the Antipodes

The European migrants – particularly the wealthier free settlers – once established in their new home, took up the outdoor recreational pursuits they enjoyed at home. Hunting, naturally, was popular for the 'landed gentry' and for the farmers as a means of supplementing their diet of mutton, beef and pork with game meats from both local animals and those imported for the purpose of hunting (such as the rabbit and the fox). City dwellers with middle-class professions, incomes and lifestyles sought rest and recreation in the nearby Australian bush. Hiking, often combined with amateur natural history studies, was a pastime for middle-class urban dwellers who had the spare time and resources to undertake it. For those who were more adventurous and had the time, other opportunities existed. Within reasonable distance from Sydney, the vast snowfields of the Australian Alps hosted skiing outings and huts were constructed to provide shelter when the inevitable blizzards occurred.

A vastly different geography to the UK and, to the new arrivals, amazing wildlife (the kangaroo, the koala, the wombat) made outdoor Australia a living museum. If a settler found the Australian 'bush' an undesirable place, even a frightening place, there was no escaping it. Land had to be cleared, houses built, bullock tracks constructed through the unforgiving scrub (which grew back at every opportunity) and over rock-strewn hills, while dodging flood-prone creeks. There was no respite from snakes, leeches and the night-time howls of dingo packs whose growl, more akin to that of the 'singing dogs' of Papua New Guinea, unnerved the British, who were familiar with the single syllable 'woof' of their domestic dogs. The early European-Australians were forced to come to know the bush – even if this was the last thing they wished to do. These people were resilient and they adapted and became different people from the pale, pock-marked and often sickly English, Irish or Scottish person they had left behind. Australia, notwithstanding its first settlers (the tribal wandering Aboriginal peoples), was a commons with a variety of game meats (kangaroos, ducks and possums) for the taking – a far healthier and more substantial diet to the one most Europeans were used to back home. Diet and the outdoors lifestyle in a warm climate had a significant impact on home-grown Australian culture. Here was the genesis of the healthy outdoors Australian, much celebrated on the world's sporting fields.

The influx of people from the mid 1850s onwards meant that vast forests were ringbarked, the dead trees dumped into large heaps and burnt to produce farming land. This was a different use of fire to that of the Aboriginal peoples, who as hunters and gatherers used 'fire-stick farming' to drive kangaroos out of the bush to be more readily hunted. There were never more than few Aboriginal peoples in any one area, while the new settlers kept coming – and land increasingly far from the towns and embryonic cities was converted to farmland. It is estimated that, by the twentieth century, only one-quarter of the original rainforests remained, the rest having been cleared for farming. First the highly valued timbers were removed then small-scale dairy farms (and mixed farms) were cut out of the forests. On the flat coastal land in sub-tropical and tropical Australia, sugar cane became the prime crop, with lowland rainforest country cleared to grow this extremely tall grass.

There was a negative side – there always is, it seems. The vast size of the country led to complacency: there would always be kangaroos, wallabies, possums, koalas and more arable land 'around the corner', more high-quality red cedar trees for house construction and, when the beauty of this timber came to be realised, for furniture making. There would always be more. The rainforests that were not completely destroyed were depleted of their most precious (read beautiful) timbers.

Yet the limits nature imposed were slowly recognised and we came to understand that the forests, the mangroves, the watercourses and the native fauna were finite. Some forests retained their biological diversity and generations of environmentalists have fought to save them from further forestry operations. Eventually the koala would be restricted to small reserves, and by the 1970s, they were no longer commonly sighted. On the other hand, the numbers of kangaroo would expand dramatically in the pastoral and farming country where previously the limited water had controlled animal populations. As the artesian bores were drilled and water spilled over the vast dry inland plains, the kangaroos benefited as much as the sheep. The lack of understanding of Australian soils (Were not all soils deep, rich and resilient like those back home?) and weather patterns (What were these monsoons and cyclones, these hundred-year droughts?) would lead to overstocking and overgrazing followed by soil erosion. It took generations for the downside of this environmental ignorance to be noticed, and even more generations before tentative steps were made to mitigate the problems.

No-one is to be blamed – the early settlers did not have knowledge or understanding of the physical geography and ecology of their new country. They were following in the footsteps of their distant ancestors, who had cleared much of the farmland in Britain and other European countries hundreds of years before (in some cases millennia before). Only time would improve the knowledge and practices of the settlers, yet economic self-interest and political interests would interfere every time an ecologically realistic solution seemed within reach. To this very day these are the real-world impediments to a sustainable future.

The first national parks

Some early settlers – a small minority at first – came to love the Australian bush for its own sake. They did not wish to conquer but instead just to hike in it from daybreak to sunset. Hiking clubs were formed; amateur ornithologists took their sketchbooks into the forests and the vast grassland plains to seek birds never seen before; landscape artists painted to their heart's content. These were the nation's first environmentalists. As more and more people engaged in these bush pursuits, political leaders came to recognise the instrumental value of preserving tracts of natural land near to the larger cities. Hard-working citizens needed respite from the factory in their one day away from work.[1]

In a climate as mostly kind as Australia's, the idea of a picnic in a rural setting was appealing. (Only in the far northern tropics was the climate not kind to the fair-skinned English, Irish and Scots.) Wandering through forests of the like that had not existed in Europe for centuries was a psychological and physical stimulus for the work-weary. The first national park to be declared in Australia was the Royal National Park just south of Sydney. It was formally recognised in 1879, not long after the first national park in the world – Yosemite, in the USA in 1864. Over 100 years ago Australia was already signalling that it was going to be an environmental trendsetter.

In the minds of the government, the Royal National Park had two equally important functions: one was to provide a recreational facility to be used by urban workers who would appreciate a day's hiking or a family picnic in their free time, and as a consequence would be more productive during the working week; the other was to protect what today we call the biodiversity of the ecosystem. It then was simply known as conservation or preservation, depending on how strict the criteria. The second national park in Australia was declared in 1908 at Mount Tamborine just south of Brisbane.

Much was to be lost

Not to give the wrong impression, this forward-looking environmental initiative (the setting aside of national parks) was to prove insignificant in the greater scheme of things. The early settlers had been determined to conquer this wild country and turn it into a productive paradise for those who had, for whatever reason, left the cold and damp of the UK. Later, others would leave northern European countries for Australia for the same reason: to escape relative poverty and make a future (if not a fortune) in a vast underpopulated land. Europeans were not the only early settlers – entrepreneurial Chinese arrived, particularly during the nineteenth-century gold rush, where they would, in due course, make their impact as market gardeners and shopkeepers. Other adventurous people came from the four corners of the world in search of gold, land, pearl shells and adventure. The hard-working free settlers came first from northern Europe, then southern Europe, because they could create their own lives. The free-spirited, whether convict or free settler, were the people to make Australia home. To this very day, Australians show in science, sport and scholarship a free-spirited and easygoing nature and a focus on success, commanding genuine respect in the world.

Much was not understood

Farming, whether grazing of animals or the growing of crops, posed serious challenges for people who knew completely different soils and had no experience of Australian predators. Animals that either preyed on or competed for food with the farmers' animals and crops were prime targets for eradication. Many native animals were declared pests, with no comprehension of their role in controlling the population of other pest animals (which, in due course, became a greater problem to the farmers than their predators). To deal with the dingos, eagles, flying foxes and anything else that ate small farm animals or fruit crops, a bounty was imposed (so much per head, or per pair of claws or pair of ears). The monetary incentive was to kill and scalp – or otherwise prove the death of the animal. Those with an instinctive understanding of economics came to realise that an income was to be made by 'harvesting' these predators at a sustainable rate, but this, of course, meant the farmers' problem remained unresolved. Before the concept of culling at a sustainable level was generally understood, enormous damage was done to populations of native Australian fauna as shooters set about making a living (they might have dreamt of a fortune). Ironically, the decimation of many native animals occurred just as the introduced rabbit and fox commenced their inevitable population explosion. With that came the destruction of land (in the case of the rabbit) and native animal populations (in the case of the fox). What rifles and traps did not accomplish in the decimation of native animals, the destruction of habitat did. The latter was gradual but inevitable as the human population increased and demand for cleared farming land kept pace.

The building of dams, the irrigation of soils that had never experienced constant water flows, as well as unintentional overstocking, gradually had an impact on the productivity of the land. Initially, as you would expect, there were gains in farm productivity. The short-term economic impacts were very positive as agriculture boomed and prices were high due to the quality of the produce. At the peak of the wool boom from the 1950s to the 1970s, Australia had 100 sheep for every human. The price of wool during the Korean War reached a pound for a pound (in the pre-1966 currency and weighting system). Where massive irrigation systems were developed (the first and most significant being in the lower Murray–Darling rivers region), fantastic food bowls were created.

These initial productivity booms were followed by slow gradual decline, more than likely unnoticed until a saltpan appeared, and we began to realise that success was not necessarily going to be permanent. In the past twenty or so years we have begun to realise that we have in many places pushed too hard. Salinity and waterlogging are serious problems resulting from over-irrigation. There were insufficient environmental flows in the lower reaches of a number of our rivers. A severe drought since 2005 has led to massive reductions in available water in the lower parts of the Murray–Darling Basin, creating a near to intractable political problem. We should not be optimistic that an environmentally sound and workable solution will be found.

Rivers and streams had their water overcommitted to farming, and as the country's population expanded (much faster than in the mature industrialised countries in Europe) water was diverted to the cities. This severely reduced environmental flows, which have done, in some cases, significant harm to downstream river environments and the estuarine-based industries, such as fishing, that rely on adequate water flows. After 150 years of population growth, the clearing of vast areas of land for farming, the ongoing development of a substantial mining industry and the spread of suburbia, Australia was ripe for an environmental revolution. It started slowly, very slowly, only becoming apparent in the early 1970s.

MODERN ENVIRONMENTALISM

Australia had a population of around 12 million people in the late 1960s. The vast majority lived in the cities that had been established in the early days of settlement. Australia's geography and settlement patterns destined it be an urbanised country – the most urbanised large country in the world with over 85 per cent of the population being urban coastal dwellers. The development of world-class industrial-scale mining and mechanised agriculture, both having extremely high labour productivity, replaced the drover, the farm hand and the pick-and-shovel miner.

What Max Nicholson calls 'the environmental revolution'[2] started in the 1960s. It had various routes in different parts of the world, from bird-watching societies, bushwalking clubs, amateur botanists and an awakening by the baby-boom generation to the fact that population pressure and urbanisation were changing forever the natural environments associated with their youth. Just as the world's indigenous peoples identified with 'their land' so the 1960s youth of modern Western societies realised they had a 'sense of place' for the environments of their formative years. The concept of a sense of place appears to be a universal human one.

The rapid pace of industrialisation starting in the 1960s was both welcomed and (mildly) feared by the baby boomers. They liked their pace of life – the surf, the freedom to wander around the bush and to swim in creeks within a few kilometres from the central business district of the major cities. They were not, in the main, 'flower-power hippies' but they did worry about the spread of 'little boxes' in suburbs, the high-rise buildings throwing shadows over their favourite beaches, the ever-increasing kilometres of 'tar and cement' and the 'rat race'.

The occasional bulk-oil carrier running aground and spilling its cargo over pristine beaches, suffocating great flocks of seabirds, helped heighten awareness that pollution could cause dramatic disasters. But unseen atmospheric pollutants from the dramatic increase in the number of motor vehicles crowding city streets and the increase in use of fossil-fuel energy went largely unnoticed until recently. It was a long time before we discovered greenhouse gases. In the 1970s, they were associated with the possibility of the next ice age. By 1987, climate warming was the threat, and has continued to be until the present – with much greater urgency today.

Until scientific experts asserted publicly and positively that the globe faced ecological problems (some would claim disasters), there was no rallying point of focus for more aware citizens. It wasn't until the late 1960s that a small number of expert voices coalesced into an environmental advocacy movement. Rachel Carson, an American chemist, who published her book, *Silent Spring,* in 1962, deserves pride of place as the catalyst for modern environmentalism. Paul Ehrlich's 1968 book, *The Population Bomb,* was the first of a number of 'doomsday' texts. In retrospect we might judge some of these early advocates to have overstated the case, but they were needed in their era to shake the complacency of the unaware. We have come to realise they touched some, a few, while completely missing the mark with those who danced to the music of Presley but didn't listen to the words of Bob Dylan. The environmental revolution started but was to be slowly drawn out, with the occasional false start, until the twenty-first century.

In the 1960s throughout the industrialised world a new generation of university students became aware of the calls to action by environmental campaigners such as Ehrlich and Carson. The awakening of the baby-boom generation was not confined to a narrow environmentalism. For example, Ralph Nader and his pro-consumer advocacy and condemnation of bodgie cars made him a hero of the 'exploited consumer', a role he continues to play to this very day. Overlaying the developing environmentalism and product safety was the 1960s 'renaissance' (or 'new Enlightenment') comprising flower power and the promotion of peace, free love, free speech, anti-racism, feminism and a new 'value' politics that was neither 'left' or 'right'. Its adherents were disgusted – to put it mildly – with both the ongoing US war on Vietnam and the 1968 Russian invasion of Czechoslovakia. Both the West and East, as conventionally defined, were out of favour. Here was the start of a new politics. The early green movement was a little-noticed part of this much bigger picture of the new Enlightenment. The Vietnam War, feminism and the (unlikely) prospect of a radical transformation of society resulting from the 1968 student revolts overshadowed the slowly emerging environmentalism. In Australia, it took the disbanding of the anti-Vietnam war campaign in 1972, when the new Labor government took Australian troops out of the war, for the environment movement to successfully compete for public attention.

Putting environmental impacts on the table

By the late 1960s trans-boundary pollution – in particular, acid rain – was inescapable in highly industrialised northern Europe and North America. Awareness of this at the governmental level led to the first-ever United Nations Conference on the Environment, held in Stockholm in 1972. Two years before, the concept of the environmental impact statement (EIS) had been established by legislation passed by the US Congress. No longer would the environmental impacts of major developments be completely ignored – at the very least lip service was to be paid to them. The EIS concept integrated the engineering, economic, environmental and social impacts (both desirable and undesirable) of major projects in a holistic way. Here was a crude precursor to the concept of sustainability and what in business is called 'the triple bottom line'. Country after country followed the US initiative and legislated for EISs. Embryonic EIS regulations came to existence in Australia in 1972 in, of all places, my home state, the then extremely conservative Queensland. By 1974, Australia had a full-blown EIS law enacted by its national parliament – evidence that Australia was not a laggard in recognising the demands of environmentalists. By and large, though, the authors of EISs were unsuccessful in achieving a holistic oversight of all issues (and most still are!) and applying the principles and analysis of all relevant disciplines.

The fight for our reefs and beaches

Going to the beach for the day, putting up a beach umbrella, spreading out the towels, buying ice-blocks or a 'tanning' spray of coconut or mutton-bird oil from wandering salesmen was as natural to Australians as was skiing to the Norwegians, Guinness to the Irish and apple pie to the Americans. A picnic box, a plastic bucket and a spade and an inflatable float – or for the 'surfers' a three-metre-long wooden surfboard – were essential items for a day at the beach. A day would become days as holiday periods extended and non-working mothers (the majority) camped in tents, while fathers arrived each Friday night after the week's work was done and returned to the city on Sunday night. The only digging up of the sand was by the younger members of the family as they built sandcastles. Then a dramatic change occurred before our very eyes.

The rapidly increasing affluence of the population after World War II, the speedy take-up of the motor car and the population growth (both natural and migrant) meant that it was not too long before the typical beach hangout in summer became crowded. The change occurred within a short twenty years, from the 1950s to the 1970s. Until the 1970s at the large number of popular seaside locations, scattered widely around the nation from Lorne in Victoria to Burleigh Heads in Queensland, extensive camping grounds of tents and caravans stood in place (rain, hail or sunshine) for the six weeks of school holidays. These 'tent cities' came to be slowly but surely replaced by high-rise apartments. As the 1970s advanced, the Gold Coast – now the second largest city in the state of Queensland[3] and larger than Hobart, the capital city of Tasmania, and a host of other Australian cities – was destined to become a sea of high rise. The turning point was a day in 1956 when the first high rise, called Lennon's Broadbeach Hotel – at a majestic six stories – was opened at the then village of Broadbeach.

Outdoor enthusiasts, sports- and recreation-minded people, and nature lovers were not overly impressed with the changes being foisted on their beaches – even though it was the strong demand by fellow Australians for resort-style development that drove the tents out. Here is one of the many nuances of environmental management: people have different tastes. Even the same people might want a camping holiday one year and to stay in a luxury resort the next. Yet even dedicated resort-goers were disturbed to find the ocean views they came to enjoy were disappearing as one after another high-rise apartment narrowed the prism to the beach.

What was needed to firm up this slowly developing disquiet with the overdevelopment (as it was viewed) was an easily identified 'bad guy'. Sand-mining – the ploughing then digging up and, as technology kept improving, the large-scale dredging of coastal dunes and whole beaches for black mineral sands – was just the right target for those opposed to what was occurring on the coastal strip. The space age (Sputnik 1 was launched in 1957 and the USA spent the next decade catching up) and new paint technologies were the reason for mining the beach idyll. Rutile, zircon and ilmenite – the black sands of the beach – had become very valuable. Great piles of white silica sand (what you and I know as normal sand) was moved and processed to remove the miniscule amount of mineral (black) sands within the silica.

For Australians used to the peace and tranquillity of the beach – the only noise being waves on the sand – this was an affront. The damage caused by mining was more immediately obvious than the gradual change in urban form caused by development – only when the beaches were covered by shadows cast by the high-rise apartments did we realise what had happened. The first environmental protests were not far away. They would become acrimonious affairs, claiming victims along the way, before the green campaigners eventually won the battle for the beaches. We might have chosen the wrong target in singling out the miners – they at least attempted to return the beach to something resembling its original appearance.

The Great Barrier Reef

At about the same time that beachgoers and nature lovers (from the Myall Lakes in NSW to Fraser Island in Queensland) were mounting their campaigns against the sand-mining companies (some local Australian firms, others multinational businesses) another environmental battle was looming. Stretching from Lady Elliot Island (north of Fraser Island) in the south to Papua New Guinea in the north is the Great Barrier Reef, the largest coral assemblage in the world. Captain James Cook on his northward journey, after claiming Botany Bay for the British, ran aground on the reef east of the northernmost Queensland town that today bears his name. The reef's danger to shipping has been virtually eliminated with advances in pilotage, and its status as an iconic environmental asset (of World Heritage value) is now firmly cemented. Yet, as I write, there is considerable concern that unless climate change is brought under control the reef's future is highly uncertain.

Coral reefs are primarily limestone, an essential ingredient in cement-making. In 1967 an application to mine limestone from the Great Barrier Reef corals was the catalyst that stirred a massive nationwide campaign to protect the reef. We will never know whether the proposed limestone mine by itself would have drawn the number of protestors needed to win the battle, but the subsequent threat of oil drilling ensured the numbers were quickly to become high enough and the pro-environment feelings strong enough.

Australia is not an oil-rich country. It obtains some oil from the Bass Strait, between the mainland and Tasmania. It has rather extensive natural-gas fields off the Western Australian coast. This is it in terms of liquid and gas fossil fuels, the stuff that cars, utilities and trucks run on. The reef was thought a prospective area for oil in the late 1960s and applications were made to obtain permits to drill. Once this became publicly known, the battle to protect the reef began. Limestone mining is potentially small scale, while oil platforms were likely to be spread widely – and the threat of oil spills far reaching. The protest movement grew in numbers and became strong in the major southern cities while retaining its base in Brisbane and the Gold Coast.[4] Coastal sand-mining and the threat to the reef led to the foundation of Australia's first environmental group to focus on coastal and marine ecosystems. Its name, the Queensland Littoral Society, suggested its focus and its leaders in the early days were unsung heroes of the environmental movement.[5] At a national level the Australian Conservation Foundation did its best to support the Queensland campaigners and make the issues of national significance.

Government matters

The battle for the reef was eventually won, but not before a great deal of effort was put into convincing the Australian government that it was to its political advantage to confront the Queensland state government. The Queensland government was forced to accept a royal commission into the costs and benefits of oil drilling in reef waters. When the first Labor government in 23 years was elected at a national level in 1972, the writing was on the wall for the prospective oil miners and their backers in the Queensland government. The Labor politicians were (by inclination or political self-interest) inclined to favour the environmentalists. This is not to suggest that the Labor Party was 'green' in a political sense, yet the short three years of the Labor government, led by Gough Whitlam, resulted in a radical change in governmental attitudes to the environment.[6] The next government, led by the Liberal Party leader Malcolm Fraser, was equally pro-environment, carrying on with and implementing the reforms of the Whitlam government.[7]

In 1975 the national government enacted a law to protect the reef from all mining. In 1979, work began to develop the first zoning plan for the reef – dividing it into highly protected areas, tourist areas,[8] fishing areas and the remainder, where virtually all activities except mining were allowed. The Great Barrier Reef Marine Park (its formal title) was the largest protected marine area and the first marine park in the world when it was formally established. It retains this status. The first Marine Park Plan for the Great Barrier Reef[9] was approved by parliament and, in 1981, the reef became the first Australian area listed as a World Heritage Area by the United Nations Education, Scientific and Cultural Organization (UNESCO).

Within a matter of a few years Australia was to have more parts of the nation listed as World Heritage properties (as they are formally called) for their natural values than any other country in the world. Three facts made this possible. First, Australia is a richly biodiverse country with unique flora and fauna. Second, it has a small human population so has suffered less damage to the country's most valuable natural environments, some examples being the reef, the remnants of the rainforests in north Queensland and the large area of temperate rainforests in south-west Tasmania. Third, is the existence of a well-organised environmental movement built on a strong base, the leadership of which, in the early days, comprised superb communicators and organisers who were not afraid of the state or of vested interests.

The sand-mining battle was, for the most part, played out separately from the reef battle. Not long after the reef battle was won, the Fraser Island one started in earnest.[10] Sand-mining on the island was made subject to an official inquiry under new national legislation. The inquiry took place in 1976 and the result was the prohibition of the export of mineral sands, effectively bringing the industry to a halt on Fraser Island. Another dispute about what land uses to permit on the island arose in the 1980s; this time the focus was timber-getting. Again the conservationists won (after another public inquiry – a preferred method of settling environmental disputes in the 1970s through to the 1990s).

The fight for our rivers

Tasmania – the smallest Australian state, separated from the mainland since the last ice-age thaw and consequent sea-level rise – has managed to play a crucial role in environmental politics in Australia, far more so than we would expect given its small population and economy. When Moss Cass, then federal environment minister, introduced the bill for the adoption of environmental impact statements in Australia in 1974, he singled out the highly controversial flooding of Lake Pedder in Tasmania as a catalyst for the new law. There had been great dispute and controversy over the issue of the flooding of Lake Pedder caused by damming as part of a hydroelectricity scheme. No comprehensive assessment, as would be required after Cass's law was approved, had been undertaken. Cass suggested that, in the future, carrying out EISs would lead to more informed decision-making than had been the case with Lake Pedder.

Tasmania is not only geographically small, but has a small population and much natural land as a consequence. A large area of temperate rainforest in the south-west remains intact – it would have taken a Herculean human effort to clear forests that grow both thick and wide, spreading horizontally after reaching a couple of metres in height. This is a landform that guards nature without human help – you would have a better chance walking along the top of the rainforest than through it. However, the magnificent wild rivers that flow through this forest area present opportunities for hydropower schemes – all that is needed is a dam.

Tasmania runs its economy – and people's homes are lit and warmed – on hydropower. Hydropower requires the falling energy of water to spin the turbines and if natural rivers don't provide the drop and volume of fall, dams have to be built. Throughout the 1960s, and into the 1970s and 1980s, Tasmanian politics centred on the debate between those supporting the extension of hydropower (via the construction of new dams) and those wishing to see the wilderness in the south-west remain intact. In the early years the hydropower advocates won the debate, but the balance of power shifted to the conservationists in the 1980s when Tasmania became the centre of big 'G' Green politics in Australia.[11] In 1983, the High Court of Australia ruled against the proposal by the State of Tasmania to build a dam to flood the Gordon River. The environmentalists had won a major battle.

It is not just free-flowing water that excites developers; there is considerable natural forest in Tasmania which they eye off. If your business is selling woodchips to Japan, for example, natural forests are a free gift of nature and you don't have the costs of purchasing land, planting and nourishing the trees and waiting many years for them to reach a harvestable size. Plantation forestry requires the timber and woodchip companies to pay for the timber. When governments are willing to subsidise access to natural forests (as they often are), these timber companies celebrate their inflated share prices and profits. Running parallel with the political battles over hydropower have been battles over expanding forestry operations and pulp mills. For its size, Tasmania has been (and is) the site of more than its share of environmental disputes. These have gained international attention and the involvement of eminent foreign environmentalists such as David Bellamy. Tasmania is unique in this regard.

Green politicians

The small population of Tasmania, and the proportional representation system that is used to elect Australians to the Senate (a 'house of review' but able to exercise real power in frustrating governments if they do not have the numbers to control it), provided the ideal situation for the election of the first Australian Green politician to the national parliament. Bob Brown has used his position as senator on every opportunity – usually in coalition with the other strongly pro-green party, the Australian Democrats – to promote preservation of large tracts of land in his state. However, Brown is far from being a narrow parochial politician; he has done much to influence Australian public policy on a range of issues. Brown's politics are those of a left-liberal on non-environmental issues.

Within the Tasmanian state parliament, the local Greens have played a significant role,[12] at one stage being in formal coalition in government with the Labor Party. This arrangement did not last due to strong tensions between pro- and anti-development interests in the Labor Party, which has its electoral base in two distinct constituencies. One constituency is the workers in industries such as timber-getting and woodchipping, and hydropower, and the small businesses that are aligned with these industries. The other constituency is the urban-dwelling professionals who have their futures in the service industries, including education and ecotourism, both of considerable importance in Tasmania. In attempting to juggle conflicting political demands, the Labor Party moves from one powerbase to the other, then back again. This has left room for the state-based Greens to be successful in recent elections.[13]

The wet tropics of North Queensland

To the other end of Australia – the city of Cairns is a two-hour flight north from Brisbane, the capital of the state of Queensland. Two hours in a fast aircraft from Brisbane and you are only halfway up the Queensland coast. The tip of Australia, from which you are only a canoe trip away from Papua New Guinea, is as remote from Brisbane as Fiji, New Zealand and the Solomon Islands. Australia is a *very* large country.

From Cairns drive for two hours through sugar-cane fields and across mangrove-lined rivers, skirting the brilliant light-blue waters of the Great Barrier Reef, through the small tropical town of Mossman, and then more cane fields and you finally arrive at the Daintree River. A ferry ride of under five minutes takes you into the Daintree. Many visitors (and for that matter, Australian residents) refer to the World Heritage Area, formally called the Wet Tropics, as 'The Daintree'. The Daintree is in fact a small part of the Wet Tropics World Heritage Area and the story of how it was declared a World Heritage Area is another of the big environmental battles that has given Australia a particular place in global environmental politics.

The growth-at-any-cost mentality of the then Queensland premier, Joh Bjelke-Petersen, ensured that the battle of the rainforests would be bitter, long-running and, at the end of the day, a win for those with reason on their side. Green advocates who spent time at the frontline of the battle for the forest were harassed by police, condemned by significant segments of the local population and cheered on by those who wanted to see the Queensland premier lose.[14] You didn't have to be green to be in this group.

It took Commonwealth government intervention – a willingness to nominate the area for the World Heritage List, plus provide significant financial compensation to those who would lose jobs (timber workers in particular) – to resolve the issue in favour of the conservationists. Today, the Wet Tropics is a major setting for ecotourism in Australia.

Resolving environmental disputes

The history of environmental disputes in Australia (the Wet Tropics battle stretched to the end of the 1980s) caused a rethink of process by those who had to resolve these matters. Increasingly, the task fell to the Commonwealth government. The traditional resource-resolution model pitted the narrow agenda of economic departments against the 'wider' agenda of environment departments – each on a different plane. The need for a unifying decision-making principle was obvious. This came about in the late 1980s when the Hawke Labor government established the Resource Assessment Commission,[15] whose task was to analyse – from an economic-cum-ecological perspective – major resource-use complaints as they arose in Australia and make recommendations to the Commonwealth government on the means of solving them. The commission's recommendations were to be made public to help inform the average citizen of the utility of this approach. The commission promised much – much more than it could deliver given the inability of most people to grasp the need to marry economics and ecology and, although it did not survive the next change in party leadership,[16] it was groundbreaking – and decidedly Australian – innovation in environmental decision-making.[17] It is now forgotten, yet the law establishing it is still in the statute books. Perhaps its time will come again and the statute will be dusted off.

The commission was not the last of the significant home-grown environmental initiatives in Australia; however, it was the second last. With the change of national government in 1996, the Labor Party lost to the Liberal–National Coalition and a new environment minister, Robert Hill, a small 'l' liberal with a strong interest in the environment and progressive Enlightenment policies, was appointed. Hill took the by then old *Environment Protection (Impact of Proposals) Act 1974* and turned it into a world-leading piece of legislation, the *Environment Protection and Biodiversity Conservation Act 1999*. To give a good example of why this law deserves its elevated status consider that under the act Australian fishers have to meet the highest sustainability criteria in how they use the oceans (in terms of the level of their harvests and the protection of non-target animals) before they are allowed to export fish. How many countries have gone this far?

Robert Hill, utilising his professional legal training, was a brilliant advocate for Australia in 1997 Kyoto Climate Change talks and subsequent agreement. Of all mature industrialised countries, Australia won the most advantageous concessions at this meeting. There should have been no reason for the Australian government not to sign the Kyoto agreement –

but it didn't then.[18] Regardless of Hill's standing in his government (he was leader of the government in the Senate), Australia was denied a place in the sun as a progressive, pro-environment nation when the prime minister (and, presumably, the majority of the Cabinet) of the time refused to ratify the Kyoto Protocol. Australia and George Bush's USA would stand alone as climate change 'deniers'. Hill was 'promoted' out of the environment portfolio.

Australia is once again in a position where its green credentials are at stake. The new Rudd Labor government has before it a 'middle-range' recommendation for a greenhouse gas emissions trading scheme. This is the equivalent of a carbon tax as it would impose a cap on emissions and individual companies would be required to purchase permits at an auction and then trade them if they wished.

Australia's place in the world tends to wax and wane. It is rich. It ranks very highly on the HDI, currently third after Iceland and Norway. It has magnificent well-protected World Heritage areas and some of the best environmental legislation in the world, yet during the last part of the Howard Liberal government's tenure, the environment was downplayed. The conservatives in the Liberal Party had gained the ascendency and the main game was to support the policies of the USA. Australia is very much a work in progress.

1   This was before the half-day holiday on a Saturday. The 40-hour week was not to come until the Australian trade union movement flexed its muscle in the late 1800s.

2   Max Nicholson, *The Environmental Revolution: A Guide for the Now Masters of the World*, Hodder & Stoughton: London, 1969.

3   The Gold Coast has a population of approximately half a million people.

4   Judith Wright-McKinney, as she became known later in life, a well-known Australian poet whose works were studied in high school from the 1950s, would also be remembered as a leader of Australia's first major green campaign. She was president of the Queensland Wildlife Preservation Society.

5   The Queensland Littoral Society, founded in 1965, became the Australian Littoral Society in 1969 and is now known as the Australian Marine Conservation Society. Its early leaders were Eddie Hegerl, Di Tarte and Des Connell.

6   It is important to emphasise that the change of national government in 1972 resulted in a dramatic shift in political ideology. A Labor government held power in Australia immediately after World War II, but from 1949 the then somewhat conservative Liberal Party, in coalition with the Country Party (known as the National Party since 1982), held power until 1972. The coalition was helped to remain in power by a more right-wing breakaway group from the Labor Party, called the Democratic Labour Party, which sided with the non-Labour parties. It was Catholic and strongly anti-communist and this stance, as well as its vigorous opposition to the Labor Party, is not easy to explain as the Labor Party had nothing in common with the Communist Party of Australia from the 1920s. However, ideological differences based on religion divided the working classes of Australia for a long time. The irony – as there so often is in politics – is that the non-Labor parties supported by the Democratic Labour Party were overwhelmingly Protestant – strange bedfellows!

7   It has been an interesting feature of mainstream politics in Australia that equally powerful, forceful and successful environment ministers have been found on both sides of politics. The longest serving and, it could be argued, most dedicated and radical environment minister was the Liberal Party's Robert Hill, who served in this role from 1996 to 2001.

8   These are equivalent to terrestrial national parks under Australian law.

9   In 1979, I was honoured to be relieved from my university duties so I could become, for a period of twelve months, the Marine Park Planning Officer and develop the first Marine Park Plan for the Great Barrier Reef.

10   John Sinclair, a bulldog of a man with a conservative rural background, led the sand mining campaign on Fraser Island in Queensland from the front and took the brunt of the flack from the pro-development Queensland government and the mining industry. He lost his job, was sued by the then Queensland premier (Joh Bjelke-Petersen) and was eventually forced – for his sanity – to leave Queensland to take up permanent residence in Sydney.

11   There is hardly an Australian who doesn't know of Senator Bob Brown, the nation's first Green elected to the federal parliament.

12   It was as early as 1972 that Dr Dick Jones, a University of Tasmania academic, sought election under the banner of the United Tasmania Group (the UTG, a forerunner to the Greens) and went within a few votes of being elected. His legacy – although indirect and distant in time – today is that a great swathe of the south-west of Tasmania is a UNESCO-listed World Heritage Area. It is truly a magnificent part of Australia.

13 The particular electorate system in Tasmania (unique in Australia) is the Hare–Clarke proportional representation system.

14 Dr Alia Keito, an unknown scientist until then, made her name as the person who dedicated years of her life to detailing the biological diversity of the North Queensland ancient forests. Activists such as Mike Berwick, who went on to become mayor of Port Douglas, the local shire, and Rosemary Hill, who remains an active environmental advocate, are two of the better known green advocates in the region.

15 A colleague of mine, John Ware (economist and avocado farmer), discussed with me the idea of writing to John Kerin, the key resource minister in the Commonwealth government, to propose the establishment of a truly interdisciplinary (ideally, trans disciplinary) resource assessment body. Not longer after, the commission was established.

16 For a range of reasons, some to do with the inability to firmly establish an integrated (economic–ecological) analytical methodology and some to do with personal politics – when Paul Keating replaced Bob Hawke as prime minister, initiatives with Hawke's handprints on them tended to disappear – the body was let wither on the vine after a couple of years.

17 The only equivalent innovation was the position of environment commissioner with the Industry Commission, a position I held as the inaugural appointment from 1990 to 1996.

18 The Labor government, under Prime Minister Kevin Rudd, signed the instrument of ratification of the Kyoto Protocol in December 2007 and it came into effect in March 2008.

CONCLUSION

It is the nature of the human species to learn from mistakes. How many of our ancestors died after eating a poisonous fruit as they surely but gradually discovered what is edible and what is not? We don't know the answer to this; however, in the twenty-first century we can feast upon a great range of delicious fruits – the same applies to vegetables – because we are inquisitive and capable of learning. Sophisticated language and our ability to make records, recall the past and orient the future to suit our ends distinguishes us from other animals.

Each chapter in this book recounts events, analyses them and teaches us something. The lesson might be as simple and sensible as the proposition that wars can be waged without injury or death – as in the cod wars between Iceland and the United Kingdom. There are complex lessons: for example, we know that the old Soviet Union was an environmental vandal because the gifts of nature did not feature in Soviet-style economics; yet there is no simple answer as to why after the fall of the Berlin wall Russia would make a real economic, political and environmental mess in its transition to a form of capitalism. There are lessons still in the making: incomplete as we don't know the end of the story. For example, will the United Arab Emirates (and its novelty cities of Dubai and Abu Dhabi) sustain a healthy and wealthy population after oil? Unfolding stories, like our lives, are the norm.

Fortunately, we can write the future chapters, and we can write them so as to fulfil an enduring human dream for peace, prosperity and a good sustainable life. This endeavour will require great effort – by each and every nation state, by the United Nations and by individuals. The majority of human beings are capable of making significant efforts. Some humans make extreme efforts. In our quest for peace and prosperity, and in sustaining these much-desired attributes, each and every one of us will need to make an intellectual effort, learn from past mistakes and take notice of successes and move forward. This book sets out to instil an attitude of inquiry and action.

Making progress will require using our political skills to build coalitions of like-minded people so as to revolutionise our governments to make our economic system reflect the costs of pollution and the depletion of resources and to develop piece by piece, state by state, such social and economic institutions that create the social capital that is so scarce today.

We must be realistic. History indicates that only with the threat of major life-threatening and property-destroying disasters will humans put aside religious, ethnic and class divides to fight the common enemy. Until now, the level of threat posed by environmental degradation and destruction has not been serious enough to stir us from complacency. We have not viewed pollution, loss of biodiversity or soil degradation as a global problem requiring a global solution. Maybe – it is by no means certain yet – the global threat of climate change (severe weather events) will be the call to arms to unite our efforts.

As we come to the end of the first decade of the twenty-first century we have come to realise that three common enemies are bearing down upon us at rapid speed. By 2050, the world's population will be 50 per cent greater than it is today. Yet we do not adequately feed the existing numbers. We do not provide clean, safe water or the safe disposal of waste, including sanitation services, to all our present population. Given that millions upon millions of our fellow human beings live for only half the number of years those of us lucky enough to be born in the wealthy industrialised world live, we have stark evidence of the brutality of a social and economic system divorced from environmental realities.

The fact that we have the ability (the agricultural land and the other necessary resources) to feed every human alive today at an adequate level is evidence of our intellectual tardiness and lack of empathy. We recognise that feeding the world is an economic problem. While the 'deserve theory' in economics continues to hold sway – people obtain the income they deserve; they therefore get to eat what they deserve – we will starve hundreds of millions. Reverend Thomas Malthus was proved wrong year in, year out, century in, century out – until now. Malthus did the very elementary, but too easily overlooked, arithmetic that showed that the geometric growth in human population would eventually outstrip the arithmetic growth in food production. Technological revolutions have kept us ahead of the game until recently.

There are those today who doubt the validity of Malthus's mathematics. They can prove that we could feed everyone alive today. This is true – but who is going to refashion our economic system with the goal of feeding everyone as first priority! The same, somewhat naive, souls will tell you that 'trickle down' economies will solve the problem for the poor. They can, with justification, point to how the trickles from the Industrial Revolution eventually broke into strongly flowing streams with the results we see in the wealthy world today. However, if we are to wait for the trickles of the capitalist economies of China and India to work their magic, we can expect another two to three generations of extreme poverty for the workers and peasants of these countries. And after this long period (50 to 75 years), we still can't guarantee the end of poverty. And this is for good reasons.

Malthus obviously could not foresee the discovery of oil, the mechanisation of farming, the formulation of artificial fertilisers or the recent 'green revolution'. However, he lived recently enough to recognise that the world could be circumnavigated in a few weeks of sailing and that much of the world was vast oceans, arid deserts and bitterly cold tundra. Agricultural land was limited. Superphosphates were limited. Renewable energy sources were limited. It was not until 1972, in the bestseller *Limits to Growth*, that a Malthusian understanding of our finite planet slowly reappeared after an absence approaching 200 years. It was in 1968 that Paul Ehrlich had explored the threat of the human population outstripping nature's life-support system and invented the 'population bomb', which was as potentially dangerous as the atomic and hydrogen bombs. Notwithstanding the fact that the serious threats of overpopulation and overconsumption are now respectable arguments for those who care for our future (who doesn't?), the criticism levelled against the twentieth-century Malthusians delayed by a generation or more a widespread concern for the environment.

Those who deny reality are, not surprisingly, without a clear view of it. Unless we look behind the economic statistics, we are tempted to optimism and an anti-Malthusian stand. The slow degradation of agricultural lands has been masked by ever-increasing applications of fertilisers. Agricultural and mineral productivity has only grown because we have used ever-more petrol, diesel and coal to power machines. This is not going to last forever. Those who took notice – and cared – knew of Hubbert's 'peak' for oil production from the 1960s. Recognition that all finite resources were subject to a Hubbert's peak slowly took hold, and now the scramble is on to find replacement fuels. There are those supreme optimists and economists who see the numbers on their computers but not what is behind these figures, who will argue that there is not a scarcity problem because the price of mineral resources has decreased over time – and when scarcity does appear it will automatically – trust them, they say – lead to new discoveries of energy sources and new technologies.

Water scarcity is a reality today in the driest parts of the world. That enigma of a country the UAE would regress to its pre-oil days, with wandering tribes eking out a living on dates and few fish, if not for desalination. Without oil it would not have fresh water. Nature gives and nature takes. China is building dam after dam to provide irrigation and hydropower for its present population. What happens when its population doubles as the one-child policy breaks down in the face of demands from an increasingly powerful non-communist middle class? Keep in mind that at least one-third of the Chinese population lives in rural poverty. The trickle-down from the Shanghai billionaires might feed the hard-working urban poor – the trickles do not find their way to the rural folk.

The irony is that as the world – at long last – faces Malthusian limits we are the most wasteful generation of all time. Cheap fossil fuel and its derivatives have masked the reality of a finite globe. Before artificial fertilisers, animal and human wastes provided the necessities of sustainable agriculture. There was a natural cycle at play.

Before massive double- and triple-rigged transport trucks and giant container ships, renewable energy moved our goods: horses, oxen, camels on land and wind at sea. The desire to trade (to experience different foods, clothing, materials, precious jewellery and whatever was on offer – including slaves) opened the far-flung extremities of the globe to each other and, more importantly, brought new ideas. Because pack animals had to be fed and replaced as they invariably aged, only the most precious goods were transported long distances: scents, herbs, precious stones, silk and seeds. Obviously slaves were very valuable given that they were transported across vast oceans.

Today we waste enormous amounts of energy – and produce unnecessary greenhouse gases – in the name of free trade. Don't get me wrong: the idea of free trade is good; the idea of fair trade is even better. Fair trade is free trade between equals. As I was completing this book, the Australian government announced that it was likely to allow the import of bananas from the Philippines, a few thousand kilometres away. Even the most efficient container ships produce greenhouse gases. We might be able to justify this trade if we knew that the peasant farmers of the Philippines were to benefit. One of the barriers to alleviating poverty in the poor countries is the existence of subsidies to rich country farmers, subsidies that allow the latter to compete on price terms. But we don't expect that poor Filipinos will benefit in this case. In fact, we know that of all the South-East Asian countries the Philippines has been crying out for land reform for the longest. Give poor farmers appropriate-sized farms and let them reap the economies of scale of the large plantations and the case is likely to be different. It won't happen.

What makes the importation of bananas into Australia nonsense is that we discard – waste – about 20 per cent of the home-grown crop because the fruit is not aesthetically appealing to supermarket chains. They won't take single bananas or 'doubles' (twin bananas) or ones with the tiniest skin markings. The supermarket chains justify this by asserting – and it is purely an assertion – that this is what the consumer wants. Presumably, the consumer wants cheap bananas from the Philippines while ours rot on the ground. There are numerous other really silly food waste stories. However, note that not all the popular food miles tales are true cases of waste, as it does make sense to produce food where it is the least environmentally and economically expensive to grow and transport it as long as the transport costs, including greenhouse gases, are factored in. We are talking about a total (or life-cycle) approach here.

Overall in the environmental field we are slowly learning from our mistakes, yet when it comes to waste and misused opportunities the journey is proving to be very slow – two steps forward, one step backward, if we are lucky.

Take the climate change debate. Should we 'cap and trade' or tax carbon dioxide? How do we bring the developing countries into a global scheme? All agree, including them, that a global solution is the only answer. We blame the poorer countries for the potential disasters of severe climate change while we fill our homes with inexpensive goods manufactured in the same countries. This does not mean we should close our doors to their goods. Those of us who care about the plight of our fellow humans are delighted that the poor in China, India and a few other countries are climbing out of poverty due to the fact that we like their cheap products, but we do virtually nothing to help them leapfrog over the existing dirty manufacturing technology. Some of us, particularly Australians, like the fact that the Chinese purchase our rather dirty coal so that they can fire their power stations to produce the electricity that drives their sewing machines that stitch together our jeans. Once you start to analyse it in these terms, the complexity, and the difficulties, of making a better, more environmentally healthy world become obvious.

Yet there is much we could do now. The enormous 'no regrets' gains to be had in energy efficiency, in converting to non-polluting and renewable energy sources, escape our attention. When your next light bulb ceases to work, replace it with a long-life, super-efficient one. This is the very first step on the way to a completely new lifestyle. Our existing technology, architecture and engineering are of such a standard that, if we wanted to, we could construct and live in dwellings that produce no carbon dioxide. The same applies to our office buildings. My office at Bond University on the Gold Coast is in one such building. Why don't we (obviously gradually, because we are not going to demolish existing buildings helter-skelter) convert to this style of living and working? The answer is that we are not properly analysing what we do. We are not doing the economic sums. 'No regrets' means that if we do something environmentally sound and sensible – and a major threat such as climate change does not prove as dangerous as we thought – we have lost nothing. The thing we did – change the new light bulb – was worth doing in terms of money saved, regardless.

Not only are we slow in learning environmental lessons, but also we are just as slow in our willingness to try new beneficial economic and social arrangements. We have underway a few fascinating experiments in these fields. Earlier chapters have dealt with the situation in a number of countries, and I have been bold enough to make comparisons. If we are willing to open our eyes, and discard a few political prejudices, we will see that a beneficial evolution in public policy has been underway in Nordic countries for some time now. My study abroad students from these countries are bright, happy and good people. They are very fortunate to be from countries where scholarship, free education and social capital are ranked very highly. The proof, as is said in Australia, is in the pudding.

We humans are capable of brilliant ideas and inventions – in modern medicine, electronics and information technology to name the most obvious, with some noted exceptions – but we have yet to learn the lessons that will prove that we deserve our place on the planet as the thinking animal.

INDEX

b = box, *t* = table